BIOLOGY FOR ENGINEERS

AUTHOR

DR. ANNAMMA ODANETH (PhD)

Associate Professor
DBT-ICT Centre for Energy Biosciences
Institute of Chemical Technology
Nathalal Parikh Marg
Matunga, Maharashtra (India)

REVIEWER

DR. N. RAVI SUNDARESAN

Associate Professor
Department of Microbiology and Cell Biology
Indian Institute of Science, Bengaluru
Karnataka (India)

KHANNA BOOK PUBLISHING CO. (P) Ltd.
4C/4344, Ansari Road, Darya Ganj, New Delhi, India

All India Council for Technical Education
Nelson Mandela Marg, Vasant Kunj, New Delhi, India

Dear Readers,

To prevent the piracy, this book is secured with HIGH SECURITY HOLOGRAM on the front title cover. In case you don't find the hologram on the front cover title, please write us to at contact@khannabooks.com or whatsapp us at +91-99109 09320 and avail special gift voucher for yourself.

Specimen of Hologram on front Cover title:

Moreover, there is a SPECIAL DISCOUNT COUPON for you with EVERY HOLOGRAM.

How to avail this SPECIAL DISCOUNT:

Step 1: Scratch the hologram

Step 2: Under the scratch area, your "coupon code" is available

Step 3: Logon to www.khannabooks.com

Step 4: Use your "coupon code" in the shopping cart and get your copy at a special discount

Step 5: Enjoy your reading!

ISBN: 978-93-5538-698-4

Book Code: UG409EN

Biology for Engineers

by Annnamma Odaneth

First Edition: 2026

Originally Published by:

All India Council for Technical Education, Ministry of Education, GOI

Exclusively Published & Distributed by:

Khanna Book Publishing Co. (P) Ltd.

4C/4344, Ansari Road, Daryaganj,
New Delhi (INDIA)

Visit us at: **www.khannabooks.com**
Write us at: **contact@khannabooks.com**

To view complete list of books, Please scan the QR Code:

KPH

Printed in Bharat.

प्रो. टी. जी. सीताराम
अध्यक्ष
Prof. T. G. Sitharam
Chairman

सत्यमेव जयते

आज़ादी का अमृत महोत्सव

अखिल भारतीय तकनीकी शिक्षा परिषद्
(भारत सरकार का एक सांविधिक निकाय)
(शिक्षा मंत्रालय, भारत सरकार)
नेल्सन मंडेला मार्ग, वसंत कुंज, नई दिल्ली–110070
दूरभाष : 011–26131498
ई–मेल : chairman@aicte-india.org

ALL INDIA COUNCIL FOR TECHNICAL EDUCATION
(A STATUTORY BODY OF THE GOVT. OF INDIA)
(Ministry of Education, Govt. of India)
Nelson Mandela Marg, Vasant Kunj, New Delhi-110070
Phone : 011-26131498
E-mail : chairman@aicte-india.org

FOREWORD

Engineers are the backbone of any modern society. They are the ones responsible for the marvels as well as the improved quality of life across the world. Engineers have driven humanity towards greater heights in a more evolved and unprecedented manner.

The All India Council for Technical Education (AICTE), have spared no efforts towards the strengthening of the technical education in the country. AICTE is always committed towards promoting quality Technical Education to make India a modern developed nation emphasizing on the overall welfare of mankind.

An array of initiatives has been taken by AICTE in last decade which have been accelerated now by the National Education Policy (NEP) 2020. The implementation of NEP under the visionary leadership of Hon'ble Prime Minister of India envisages the provision for education in regional languages to all, thereby ensuring that every graduate becomes competent enough and is in a position to contribute towards the national growth and development through innovation & entrepreneurship.

One of the spheres where AICTE had been relentlessly working since past couple of years is providing high quality original technical contents at Under Graduate & Diploma level prepared and translated by eminent educators in various Indian languages to its aspirants. For students pursuing 2nd year of their Engineering education, AICTE has identified 87 books, which shall be translated into 12 Indian languages - Hindi, Tamil, Gujarati, Odia, Bengali, Kannada, Urdu, Punjabi, Telugu, Marathi, Assamese & Malayalam. In addition to the English medium, books in different Indian Languages are going to support the students to understand the concepts in their respective mother tongue.

On behalf of AICTE, I express sincere gratitude to all distinguished authors, reviewers and translators from the renowned institutions of high repute for their admirable contribution in a record span of time.

AICTE is confident that these outcomes based original contents shall help aspirants to master the subject with comprehension and greater ease.

(Prof. T. G. Sitharam)

ACKNOWLEDGEMENT

The authors are grateful to the authorities of AICTE, particularly Prof. T.G. Sitharam, Chairman; Dr. Abhay Jere, Vice-Chairman; Prof. Rajiv Kumar, Member-Secretary, Dr. Ramesh Unnikrishnan, Advisor-II and Dr. Sunil Luthra, Director, Training and Learning Bureau for their planning to publish the books on Biology for Engineers. We sincerely acknowledge the valuable contributions of the reviewer of the book N. RAVI SUNDARESAN, Associate Professor, Department of Microbiology and Cell Biology, Indian Institute of Science, Bengaluru for making it students' friendly and giving a better shape in an artistic manner.

This book is an outcome of many rounds of discussions, delibrations, contributions and suggestions of all members of my group at the Institute of Chemical Technology. We have attempted to incorporate all aspects of basic biology. The suggestions of AICTE members, experts and authors who shared their opinion and thought to further develop the engineering education in our country. It is also with great honour that I state that this book is aligned to the AICTE Model Curriculum and in line with the guidelines of National Education Policy (NEP) -2020. Towards promoting education in regional languages, this book is being translated in scheduled Indian regional languages.

Acknowledgements are due to the contributors and different workers in this field whose published books, review articles, papers, photographs, footnotes, references and other valuable information enriched us at the time of writing the book.

Author

PREFACE

"Biology for Engineers" is the result of our extensive experience teaching basic biology courses. As a bioprocess technology professor, I've always been fascinated by the intersection of biology and technology. There is an unmet the need for a comprehensive guide to biology that is understandable to non-scientists. Biology is a complicated and intimidating subject, but it is also a fascinating subject that has an impact on our daily lives. Engineers, in particular, must understand the fundamentals of biology as they intend to work on biotechnology, medical devices, and environmental sustainability projects. My guide would present topics such as genetics, cellular biology, ecology, and evolution in an engaging and easy-to-follow manner. To help readers relate to the material, I have used real-world examples and analogies. I would also include interactive elements such as quizzes and simulations to improve learning. We can foster a greater appreciation for the natural world and inspire innovation in fields ranging from medicine to renewable energy by making biology accessible to everyone.

We have included the topics recommended by AICTE in a very systematic and orderly manner throughout the book, keeping in mind the purpose of broad coverage as well as providing essential supplementary information. Attempts have been made to explain the subject's fundamental concepts as simply as possible. During the manuscript preparation process, we considered various standard text books and developed sections such as critical questions, solved and supplementary problems, and so on. By bridging the gap between biology and engineering, I hope to inspire more innovation in this field and encourage interdisciplinary collaboration. The topics are presented in a constructive manner, so that an Engineering degree prepares students to work in various sectors or in national laboratories at the cutting edge of technology. Whether you are an experienced engineer or simply interested in the potential of biotechnology, this book will provide you with valuable insights into this exciting field. If you are an engineer looking to expand your knowledge or are simply interested in the future of biotechnology, this book is for you.

We appreciate your interest in providing us with all beneficial comments and suggestions that will contribute to the improvement of the future editions of the book. Your feedback is essential in helping us make necessary changes and updates to ensure that our content remains relevant and up-to-date. We take pride in placing this book in the hands of teachers and students, knowing that it will serve as a valuable resource for their learning and growth. It was indeed a big pleasure to work on different aspects covered in the book, from research to writing, editing, and design. We hope that our efforts will translate into a positive impact on your educational journey. As we continue to evolve and improve our content, we welcome your ongoing support and feedback. Together, we can create a better learning experience for all.

Author

OUTCOME BASED EDUCATION

For the implementation of an outcome based education the first requirement is to develop an outcome based curriculum and incorporate an outcome based assessment in the education system. By going through outcome based assessments evaluators will be able to evaluate whether the students have achieved the outlined standard, specific and measurable outcomes. With the proper incorporation of outcome based education there will be a definite commitment to achieve a minimum standard for all learners without giving up at any level. At the end of the programme running with the aid of outcome based education, a student will be able to arrive at the following outcomes:

PO1. **Engineering knowledge**: Apply the knowledge of mathematics, science, engineering fundamentals, and an engineering specialization to the solution of complex engineering problems.

PO2. **Problem analysis**: Identify, formulate, review research literature, and analyze complex engineering problems reaching substantiated conclusions using first principles of mathematics, natural sciences, and engineering sciences.

PO3. **Design / development of solutions**: Design solutions for complex engineering problems and design system components or processes that meet the specified needs with appropriate consideration for the public health and safety, and the cultural, societal, and environmental considerations.

PO4. **Conduct investigations of complex problems**: Use research-based knowledge and research methods including design of experiments, analysis and interpretation of data, and synthesis of the information to provide valid conclusions.

PO5. **Modern tool usage**: Create, select, and apply appropriate techniques, resources, and modern engineering and IT tools including prediction and modeling to complex engineering activities with an understanding of the limitations.

PO6. **The engineer and society**: Apply reasoning informed by the contextual knowledge to assess societal, health, safety, legal and cultural issues and the consequent responsibilities relevant to the professional engineering practice.

PO7. **Environment and sustainability**: Understand the impact of the professional engineering solutions in societal and environmental contexts, and demonstrate the knowledge of, and need for sustainable development.

PO8. **Ethics**: Apply ethical principles and commit to professional ethics and responsibilities and norms of the engineering practice.

PO9. **Individual and team work**: Function effectively as an individual, and as a member or leader in diverse teams, and in multidisciplinary settings.

PO10. Communication: Communicate effectively on complex engineering activities with the engineering community and with society at large, such as, being able to comprehend and write effective reports and design documentation, make effective presentations, and give and receive clear instructions.

PO11. Project management and finance: Demonstrate knowledge and understanding of the engineering and management principles and apply these to one's own work, as a member and leader in a team, to manage projects and in multidisciplinary environments.

PO12. Life-long learning: Recognize the need for, and have the preparation and ability to engage in independent and life-long learning in the broadest context of technological change.

COURSE OUTCOMES

After completion of the course the students will be able to:

CO 1: Describe how biological observations of 18th Century that lead to major discoveries.

CO 2: Convey that classification per se is not what biology is all about but highlight the underlying criteria, such as morphological, biochemical and ecological

CO 3: Highlight the concepts of recessiveness and dominance during the passage of genetic material from parent to offspring

CO 4: Convey that all forms of life have the same building blocks and yet the manifestations are as diverse as one can imagine

CO 5: Classify enzymes and distinguish between different mechanisms of enzyme action.

CO 6: Identify DNA as a genetic material in the molecular basis of information transfer.

CO 7: Analyse biological processes at the reductionistic level

CO 8: Apply thermodynamic principles to biological systems

CO 9: Identify and classify microorganisms.

Course Outcomes	Expected Mapping with Programme Outcomes *(1- Weak Correlation; 2- Medium correlation; 3- Strong Correlation)*											
	PO-1	PO-2	PO-3	PO-4	PO-5	PO-6	PO-7	PO-8	PO-9	PO-10	PO-11	PO-12
CO-1	1	2	2	2	1	-	2	-	-	-	1	2
CO-2	1	1	2	1	-	-	3	-	-	-	-	-
CO-3	1	3	2	1	-	-	2	-	-	-	-	-
CO-4	1	3	3	2	1	-	2	-	-	-	-	-
CO-5	1	3	2	1	1	-	2	-	-	-	-	-
CO-6	1	3	3	2	1	-	2	-	-	-	-	-
CO-7	1	3	3	2	1	-	2	-	-	-	-	-
CO-8	1	2	3	1	-	-	2	-	-	-	-	-
CO-9	1	3	2	1	-	-	2	-	-	-	-	-

GUIDELINES FOR TEACHERS

To implement Outcome Based Education (OBE) knowledge level and skill set of the students should be enhanced. Teachers should take a major responsibility for the proper implementation of OBE. Some of the responsibilities (not limited to) for the teachers in OBE system may be as follows:

- Within reasonable constraint, they should manoeuvre time to the best advantage of all students.
- They should assess the students only upon certain defined criterion without considering any other potential ineligibility to discriminate them.
- They should try to grow the learning abilities of the students to a certain level before they leave the institute.
- They should try to ensure that all the students are equipped with the quality knowledge as well as competence after they finish their education.
- They should always encourage the students to develop their ultimate performance capabilities.
- They should facilitate and encourage group work and team work to consolidate newer approach.
- They should follow Blooms taxonomy in every part of the assessment.

Bloom's Taxonomy

Level	Teacher should Check	Student should be able to	Possible Mode of Assessment
Create	Students ability to create	Design or Create	Mini project
Evaluate	Students ability to justify	Argue or Defend	Assignment
Analyse	Students ability to distinguish	Differentiate or Distinguish	Project/Lab Methodology
Apply	Students ability to use information	Operate or Demonstrate	Technical Presentation/ Demonstration
Understand	Students ability to explain the ideas	Explain or Classify	Presentation/Seminar
Remember	Students ability to recall (or remember)	Define or Recall	Quiz

GUIDELINES FOR STUDENTS

Students should take equal responsibility for implementing the OBE. Some of the responsibilities (not limited to) for the students in OBE system are as follows:

- Students should be well aware of each UO before the start of a unit in each and every course.
- Students should be well aware of each CO before the start of the course.
- Students should be well aware of each PO before the start of the programme.
- Students should think critically and reasonably with proper reflection and action.
- Learning of the students should be connected and integrated with practical and real life consequences.
- Students should be well aware of their competency at every level of OBE.

LIST OF ABBREVIATIONS

General Terms

Abbreviations	Full form
DNA	Deoxyribonucleic acid
RNA	Ribonucleic acid
H	Hydrogen group
OH	Hydroxyl group
A	Adenine
T	Thymine
G	Guanine
C	Cytosine
Pyl	Pyrrolysine
Se	Selenocysteine
Cys	Cysteine
R	Rectus
S	Sinister
D	Dextrorotatory
L	Levorotatory
U	Uracil
mRNA	Messenger RNA
rRNA	Ribosomal RNA
tRNA	Transfer RNA
VDL	Very low-density lipoprotein
IDL	Intermediate-density lipoprotein
LDL	Low-density lipoprotein
HDL	High-density lipoprotein
SEC	Size Exclusion Chromatography
SDS	Sodium dodecyl sulfate
PAGE	Polyacrylamide gel electrophoresis
pI	Isoelectric point
BCA	Bicinchoninic Acid
HSP	Heat shock protein

Abbreviations	Full form
SARS-CoV-2	Severe Acute Respiratory Syndrome Corona Virus 2
AEC2	Angiotensin-converting enzyme 2
AMP	Adenosine monophosphate
ADP	Adenosine diphosphate
ATP	Adenosine triphosphate
C	Carbon
O	Oxygen
CO_2	Carbon dioxide
NH_3	Ammonia
NADH	Nicotinamide adenine dinucleotide
NADPH	Nicotinamide adenine dinucleotide phosphate
FAD/$FADH_2$	Flavin adenine dinucleotide
3-PGA	3-phosphoglyceric acid
TCA	Tricarboxylic acid
GDP	Guanosine diphosphate
GTP	Guanosine triphosphate
$C_6H_{12}O_6$	Glucose
H_2S	Hydrogen sulfide
H_2O	Water
O_2	Oxygen
CoA	Co- enzyme A
RuBP	Ribulose-1, 5-bisphosphate
Pi	Inorganic phosphate
G3P	Glyceraldehyde 3-phosphate
EC	Energy Charge
DAP	Dihydroxyacetone phosphate
GAP	Glyceraldehyde 3-phosphate
Cfu	Colony forming unit

LIST OF SYMBOLS

Symbols	Description	Symbols	Description
ΔG	Free energy change	e^-	Electron
$\Delta G°$	Standard Free-Energy Change	H^+	Proton
K	equilibrium constant		

LIST OF FIGURES

LIST OF TABLES

CONTENTS

UNIT 2 CLASSIFICATION.

UNIT 3 GENETICS.

UNIT 4 BIOMOLECULES.

UNIT 5 ENZYMES.

UNIT 6 INFORMATION TRANSFER

UNIT 7 MACROMOLECULAR ANALYSIS.

UNIT 8 METABOLISM

UNIT 9 MICROBIOLOGY

1 Introduction

UNIT SPECIFICS

Through this unit we have discussed the following aspects:

- *Basic of biology*
- *The fundamental differences between science and engineering.*
- *The aspect of biology as an independent scientific discipline*
- *Why do we need to study biology?*
- *Biological observations of the 18th Century lead to major discoveries.*
- *Brownian motion and the origin of thermodynamics by referring to the original observation of Robert Brown and Julius Mayor.*
- *Discoveries made in biology till today.*
- *How engineering is involved in biology*

The ***practical applications of the topics are discussed*** *for generating further curiosity and creativity as well as improving problem solving capacity.*

Besides giving a large number. ***of multiple choice questions as well as questions of short and long answer types*** *marked in two categories following lower and higher order of* ***Bloom's taxonomy****, assignments through a number of numerical problems, a list of references and suggested readings are given in the unit so that one can go through them for practice. It is important to note that for getting more information on various topics of interest some* ***QR codes*** *have been provided in different sections which can be scanned for relevant supportive knowledge.*

After the related practical, based on the content, there is a ***"Know More"*** *section. This section has been carefully designed so that the supplementary information provided in this part becomes beneficial for the users of the book. This section mainly highlights the initial activity, examples of some interesting facts, analogy, history of the development of the subject focusing the salient observations and finding, timelines starting from the development of the concerned topics up to the recent time, applications of the subject matter for our day-to-day real life or/and industrial* *applications on variety of aspects, case study related to environmental, sustainability, social and ethical issues whichever applicable, and finally inquisitiveness and curiosity topics of the unit.*

RATIONALE

Bring out the need for biology to be thought as an independent subject and highlight the fundamental importance of observations in any scientific inquiry.

PRE-REQUISITES

Biology : Basic Biology (Class IX)

Physics: Mechanics (Class XII)

UNIT OUTCOMES

List of outcomes of this unit is as follows:

U1-O1: What are the differences and similarities between biology and engineering?

U1-O2: Aspects of biology.

U1-O3: Why there is a need for biology for engineers.

U1-O4: Discoveries made in biology

U1-O5: Evolution of biology with the help of engineering.

Unit-1 Outcomes	EXPECTED MAPPING WITH COURSE OUTCOMES (1- Weak Correlation; 2- Medium correlation; 3- Strong Correlation)				
	CO-1	CO-2	CO-3	CO-4	CO-5
U1-O1	1	1	-	-	-
U1-O2	1	1	1	1	1
U1-O3	2	1	1	1	1
U1-O4	2	2	2	2	2
U1-O5	-	2	-	2	1

1.1 INTRODUCTION TO BIOLOGY

Biology is the study of all aspects of life. In many ways, the last 1000 years have seen a meteoric rise in the study of biology as natural science. For a long time, biology was thought to deal with only the classification of all known living organisms, animal behaviour, and habitats. So, in short, biology is the detailed study of living organisms. This includes the genetic, chemical, physical, ecological and evolutionary aspect of life, in general. The diverse life forms on earth are governed by a few basic facts of life, enlisted as under:

1.1.1 The complexity of biological systems

Biological systems have multiple layers on the basis of which life is based on. The intial steps in biology (and other complex sciences) are about identifying, classifying, and describing phenomena. Before delving into explanations for how a biological phenomenon works, it is critical first to describe its characteristics, structure, function, and behaviour. As a result, before the concepts of evolution can be worked out, the classification and associated morphology of various organisms must be considered. The nature of organic chemistry and its interactions to form biological molecules, which eventually assemble to form biological systems whose functioning is based on chemical principles, must be clearly understood. For example, in order to explain and understand why organisms behave the way they do, a detailed classification of life and the organization of molecular elements is required. Furthermore, the complex nature of an ecosystem is better understood if the levels of life are documented using the hierarchy that exists from atom to cell to organism to ecosystem (as in Fig. 1.1). This refers to how organisms evolved, not how biology developed. All living organisms are connected by a chain or web of life forms that affect the present. Changes in the past affect biological evolution like geology, but not chemistry, physics, or math. Biology's mechanisms are similar to chemistry, physics, and math. The properties of living organisms and their relationships to their environments and each other depend on their predecessors' past. Evolutionary processes help "explain" how an organism solves a biological challenge and improves itself to survive and sustain its existence.

1.1.2 Biology looks for a mechanism

Biology is much more than just "What is life?" It is also concerned with "How does it work?" On one level, we examine the organs and pieces of an animal or a cell to determine their function in the body. However, biology is also involved at the atomic and molecular level, working out the biochemistry of genes and proteins using the tools of chemistry and physics (as well as math). Such quantitative assessments can now be performed on thousands of genes or proteins in an organism simultaneously. This has spawned a new branch of inquiry known as "Systems Biology," which seeks to discover mechanisms in these massive datasets and describe how hundreds or millions of components interact in a biological system, such as a cell, organism, or population.

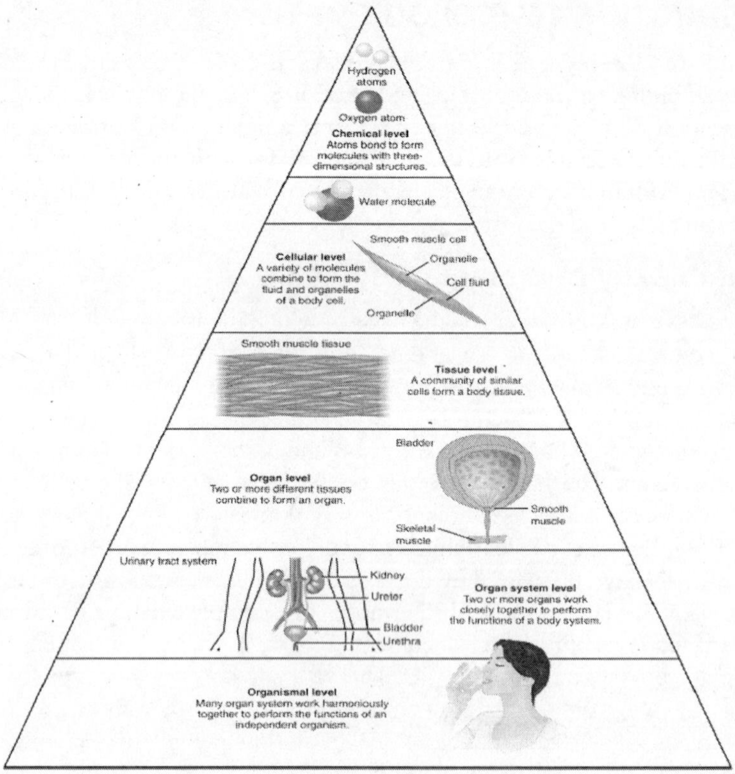

Fig 1. 1 Level of organization

Commons.wikimedia.org (Creative commons licenses)

1.1.3 Biology is multi-scaled

An organism can be studied at a variety of levels, from the atomic and molecular (biochemistry) to the internal structure and functioning of its organs and parts (physiology) to its place in a much larger system across space (ecology) and time (geology). There are two ways to approach the relationship between these scales: reductionism (looking at the smallest possible scale to find an explanation) and emergence (seeing new phenomena emerge as one looks at larger and larger scales).

1.1.4 Biology is integrative

Biological phenomena emerge from and must be consistent with chemistry, physics, and math principles. In other words, chemistry and physics govern how an organism can act or evolve. Therefore biologists must grasp how physics and chemistry exhibit themselves in biological organisms and higher-order systems. Increasingly, biologists exploring explanations of complicated natural behaviour are finding it beneficial to use mathematical, physical, and chemical models in their studies.

1.2 FUNDAMENTAL SIMILARITIES AND DIFFERENCES

The natural and physical world are the subjects of investigation that make up the corpus of knowledge known as science. The use of knowledge to design, construct and manage a product or a process that addresses a problem and satisfies a demand is what engineers refer to as "engineering" (i.e., a technology).

A scientific technique is one that is utilized by scientists. The engineering design process is one that engineers employ. A scientist begins by posing a question to the audience. The next step is for them to perform some preliminary research, establish a hypothesis, put that hypothesis to the test through an experiment, evaluate the data, and then report their findings. Engineers begin by describing the problem at hand and then determine the criteria and restrictions, generate ideas, plan, develop technology, and make improvements to their original design. Different pursuits guide the work of scientists and engineers. Scientists aim to explain and comprehend the natural world. Engineers consider a wide range of criteria and limitations when developing solutions to problems, requirements, and preferences to improve the quality of life for people, animals, and the environment.

Scientists learn through conducting controlled experiments and long-term observational investigations. The ultimate output may be a study article or a book, and its knowledge can help us understand and forecast the natural world. Engineers create things by using what they know about science. Engineers and scientists are vital in this regard, and each area benefits from the other's innovation and hard work. In certain circumstances, scientists rely on engineering advancements to improve their study (for example, microscopes or monitors). It is critical to assist pupils in comprehending and respecting the differences across the professions to help them realize the various STEM careers and possibilities available to them. A virologist, for example, is a scientist who investigates how viruses travel and how they affect the human body. A biomedical engineer can create a medication that prevents a particular virus from spreading to new cells in the body using the virologist's findings.

Even The camera and the eye have more in common than just a similar way of thinking (Fig. 1.2). They both take pictures. The camera's cornea is like a lens, and the retina is like a film. Because of these similarities, the camera looks like it has robot eyes. Even though cameras and eyes have a lot in common, they are not the same. The cornea is the "cap" of the eye. This spherical, see-through structure sits in front of the eye. The lens is made of clear glass and sits in front of a camera. Like corneas, lenses are round. Because the cornea and lens are curved, the eye and camera can only see a small area to the right and left. Without the curve, the eye and camera could only see what's straight ahead. The camera's aperture is similar to the eye's iris, which is just one of many ways the two are alike. The size of the camera's aperture controls how much light gets to the camera's sensor or film. When the iris of the eye gets smaller, the pupil gets smaller and less light gets in. When it's dark, the iris opens, letting more light into the pupil. Apertures that are wider (lower) let in more light than apertures that are narrower (higher). The less light a lens lets in, the more narrow its pupil. The eye and the camera can focus on one thing and blur everything else, no matter how close or far away it is. The eye can focus on a bigger picture, just like a camera with a wider depth of field can take a picture of a big scene. As the eye, the camera has a small field of view. The eye and the lens are shaped in a way that lets them see in the periphery. The scope of the eye is fixed, but lenses can change the camera's scope. To make a picture, the retina takes in light from the environment. The sensors in both film and digital cameras do the same thing. This is how both cameras and eyes work.

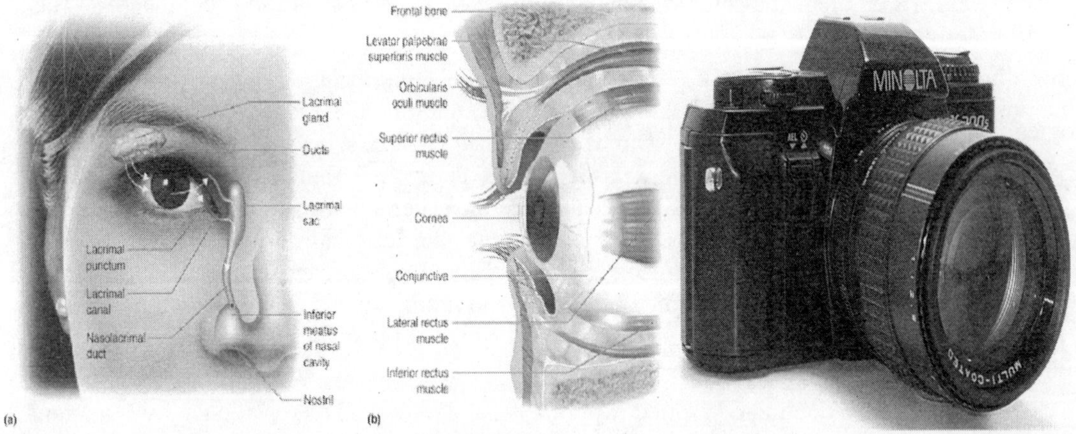

Fig 1. 2 Eye and the camera – Similarities and differences in structure and function

> Do you know?
>
> Ken Hibbard, NASA Mission Systems Engineer, shows the one-quarter-scale 3D-printed model of the quadcopter drone named Dragonfly that will land on Titan in 2034.
>
> Yet the principles behind dragonfly drones are solid. In fact, NASA has settled on a nuclear-powered autonomous craft called Dragonfly to probe the surface of Saturn's moon Titan in 2034. NASA's project is actually a quadcopter rather than a winged drone, but engineers are still convinced they can crack the code of nature's most gifted flying insect and revolutionize unmanned flight along the way.

1.3 ASPECT OF BIOLOGY AS AN INDEPENDENT SCIENTIFIC DISCIPLINE

Biology is a natural science concerned with the study of life composed of many specialized disciplines that study the structure-function growth distribution evolution, and other features of living organisms. However, despite the broad scope of biology, certain general underlying unifying concepts govern all study and research.

a. Cell is the basic unit of life

b. Genes are the basic unit of heredity

c. Evolution accounts for the unity and diversity seen among organisms

d. All organisms survive by consuming and transforming energy

e. All organisms maintain a stable internal environment

All these concepts of life reflect the impossibility of interpreting it as a physicochemical process on physical and chemical concepts. A unique structure and action relationship drive the organic world. Life can be destroyed by disruption of its molecular form; without destruction, a living structure cannot be separated from its surrounding environment or prevented from reproducing. The organic and the inorganic part of the environment are inseparable from an organism, when we study the biology of an organism, the connections with the surrounding micro and macro-environments are also essential.

Biological research indicates the first form of life on earth to be microorganisms that existed for billions of years before the evolution of larger organisms. The mammals, birds, and flowers are all relatively recent, originating within the last 200 million years. The first modern appearing humans (*Homo sapiens*) are also relatively new species, having inhabited this planet for only the last 200,000 years.

1.4 PHILOSOPHY OF BIOLOGY

Over the past few decades, there has been a rise in philosophical interest in biology, which reflects the increasing popularity of the biological sciences over the same time period. Because there is now such a large amount of written material on a wide variety of biological subjects, it is difficult to provide a condensed version of this research in a single article like this one. Instead, the purpose of this book is to explain what is meant by "philosophy of biology." Why should philosophers care about biology, and why should biologists care about philosophy? At the end of this topic, a list of the entries in the encyclopaedia that discuss certain issues in the philosophical underpinnings of biology is provided for convenience.

Three different kinds of philosophical inquiry fall under the general heading of the philosophy of biology. First, general theses in the philosophy of science are addressed in the context of biology. Second, conceptual problems within biology itself are subjected to philosophical analysis. Third, appeals to biology are made in discussions of traditional philosophical questions.

1.5 HISTORY OF BIOLOGY

Biology has been around since the beginning of time, with Aristotle, and Galen from the Greco-Roman era being the first people to study biology. Since the advent of Ayurveda, an Indian natural system of medicine, with origins more than 3,000 years ago, Muslim doctors and scholars like Avicenna have improved this old text. During the Renaissance and early modern times in Europe, empiricism and the discovery of new creatures changed how people thought about biology. In physiology, Vesalius and Harvey used experiments and careful observation, while naturalists like Linnaeus and Buffon started putting life, fossils, and the growth and behaviour of organisms into groups. Antonie van Leeuwenhoek's study of microbes with a microscope helped build the foundation for cell theory. Natural history was supported by the growth of natural theology, which was partly a response to mechanical philosophy (although it entrenched the argument from design).

In the 18th and 19th centuries, botany and zoology were essential fields of study (Fig. 1.3). Lavoisier and others used physics and chemistry to connect the living world with the non-living world. Alexander von Humboldt looked at how species interact with their environment and how this depends on where they live. He did this to lay the foundations for biogeography, ecology, and ethology. Naturalists didn't believe in essentialism and thought about things like extinction and how species can change. The cell hypothesis changed the basics of life. Charles Darwin's theory of evolution by natural selection combines these findings with embryology and palaeontology. At the end of the 19th century, the germ theory of illness took the place of spontaneous generation, but the inheritance was still a mystery.

Early in the 20th century, Thomas Hunt Morgan and his students worked on the concepts of genetics using the Drosophila model which led to the Chromosome Theory. After Watson and Crick's theory about the structure of DNA, new fields of study grew quickly. Biology was split into organismal, cellular, and molecular biology after the Central Dogma and the genetic code were broken. By the end of the 20th century, new fields like genomics and proteomics turned this trend around. Organismal biologists started to use molecular tools, and molecular and cell biologists began to study how genes interact with their environment and the genetics of natural populations of organisms.

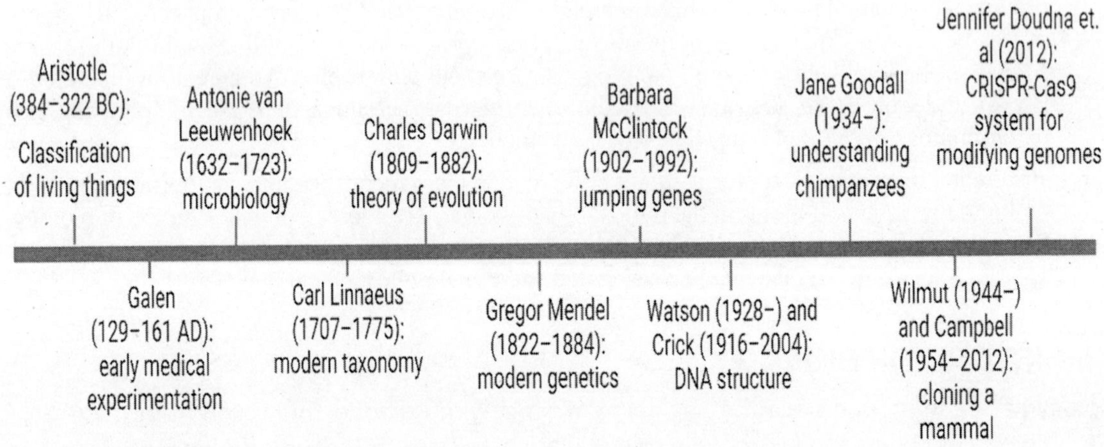

Fig 1. 3 Timeline for important discoveries in biology

1.6 NEED TO STUDY BIOLOGY

The study of biology may be found virtually everywhere and in every aspect. You are a part of biology since you are a living entity. Therefore, studying biology is necessary if one is interested in understanding the mechanisms of the human body and those of every other living entity. It is the most effective and genuine method to comprehend the planet around you.

The structure, function, growth, origin, evolution, and dispersion of living creatures are the primary topics of investigation for biologists. Along with chemistry and physics, biology is one of the major fields that provides the foundation for everything we know about the natural sciences.

A useful framework for organizing the many subfields of biology is to divide the subject into the following four categories:

- "Biochemistry" refers to the study of the chemical reactions that take place in live things or are connected to living things
- The interaction of organisms with their respective environments is known as ecology
- Study of how genes are passed down from parents to kids, as well as how these genes differ from person to person, referred to as Genetics
- Physiology is the study of biological processes, such as how a particular organ operates, what its role is, and how outside stimuli impact it,

Cell biology, environmental biology, evolutionary biology, marine biology, molecular biology, and medical biology are other important subfields that fall under the umbrella of biology.

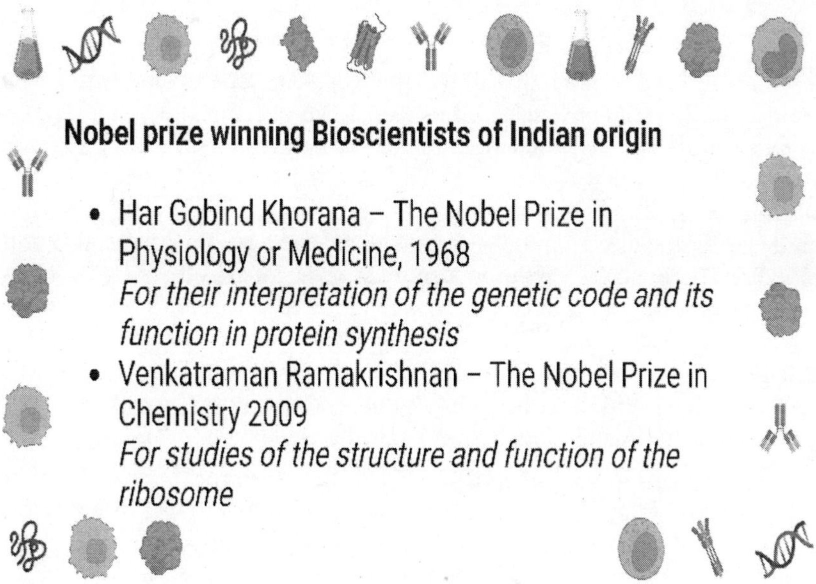

Fig 1. 4 Information table

1.7 OBSERVATIONS THAT LEAD TO IMPORTANT DISCOVERIES IN BIOLOGY

The Scottish botanist Robert Brown discovered Brownian Motion in 1827 while looking through a microscope at pollen grains hanging in the water. He was interested in the specific method by which pollen grains fertilize the female ovule. Brown observed that certain plants have oblong pollen grains rather than spherical ones. He reasoned that the grains' unusual structure would allow him to follow them and determine their role in the impregnation process. His observations of pollen grains moving randomly in water led to the term "Brownian motion," which refers to the random movement of particles in a fluid caused by collisions with other atoms or molecules.

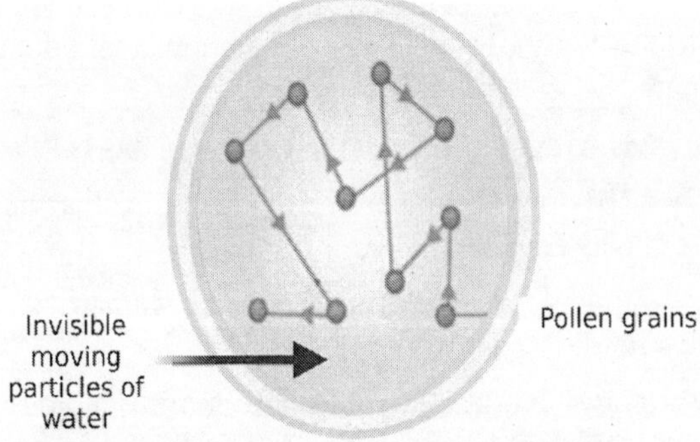

Fig 1. 5 Representation of the Brownian motion

(Images created using BioRender® software)

Thermodynamics is a physics discipline that deals with a system's energy and work. After completing his studies in 1842, German physician Julius Robert Mayer embarked on a voyage to Jakarta as the ship's physician aboard the Dutch three-masted sailing vessel Java. Mayer is supposed to have observed that storm-tossed waves are warmer than the calm sea while aboard a ship, which sparked his interest in physical principles. His questions were about the physical phenomena of warmth and whether the directly produced heat alone (the heat of combustion) or the sum of the direct and indirect amounts of heat produced must be accounted for in the burning process. He also observed that the variation in colour between arterial and venous blood is lower in tropical climates than in temperate ones. He reasoned that at greater temperatures, the human body generates less energy via burning, which led him to the notion of the equivalence between physical effort and heat. In 1842, he finally published his ideas, proposing the general rule of energy conservation and calculating the mechanical equivalent of heat.

The two instances just presented are obvious evidence that scientific observations are essential to comprehending and explaining the scientific laws that govern the science of life. Crucial components of a scientific process include making observations, keeping a record of reactions and activities within a system, and applying a scientific perspective to these components to understand the causes of the phenomena being studied.

UNIT SUMMARY

Biology is the study of living organisms and vital processes. The study of biology is fragmented; however, all branches are united by fundamental principles. Although it is traditional to separate the study of plants (botany) from that of animals (zoology) and structure (morphology) from function (physiology), all living organisms share some biological processes, such as cell division, genetic transfer, and reproduction. In molecular biology, life is considered a manifestation of chemical and energy processes. Typically, biology is organized into fundamental living units. The physicochemical properties of life are examined. Modern interdisciplinary research and the unification of scientific knowledge and investigation have led to substantial overlap between biology and other scientific fields. The disciplines of biochemistry, biomedicine, and biophysics connect chemistry, medicine, and physics to biology. With more sophisticated and accurate laboratory tools and methodologies, it is feasible to comprehend and precisely characterize the ultimate physiochemical organization (ultrastructure) of living matter molecules and how living matter reproduces at the molecular level. Biology helps us comprehend life's essentials. This attribution motivates us to apply the same principles to many gadgets to suit our demands. Scientists have made significant strides in daily operations by monitoring nature and transforming it into usable units. This chapter describes the historical discoveries that have allowed biological science to advance.

EXERCISES

Multiple Choice Questions

1) ________ discovered Brownian motion.

   A. Jennifer Doudna and Richard Roblin
   B. Justin Bieber
   C. Julius Robert Mayer
   D. Robert Brown

2) ______ is the most common prokaryotic model organism in Science.

   A. C. elegans
   B. Salmonella typhi
   C. Arabidopsis thaliana
   D. Escherichia coli

3) Biochemistry is the ______________________________.

   A. Study of the chemical reactions that take place in live things
   B. Assembly and interaction of Living things
   C. Connection of living things
   D. The systematic arrangement of living things

4) Julius Robert Mayer proposed the general principle of __________

   A. Conservation of energy
   B. PCR
   C. Electrophoresis
   D. Genetics

5) _______ discovered theory of evolution.

   A. Alvadore Luria

B. Jean Weigle
C. Charles Darwin
D. Giuseppe Bertani

Answers : 1) B; 2) D; 3) A; 4) A; 5) C

Short Answer Type Questions

1) What is Brownian motion?
2) Write about the significance of biology.
3) Draw similarities between eye and camera functions

Long Answer Type Questions

1) Give similarities and differences between biology and engineering.
2) Briefly write about discoveries in biology.
3) What are the major aspects of biology?

KNOW MORE

- https://bio.libretexts.org/Bookshelves/Introductory_and_General_Biology/
- http://mentalfloss.com/article/91177/how-our-eyes-see-everything-upside-down
- https://www.sciencelearn.org.nz/resources/8-the-role-of-observation-in-science

REFERENCES AND SUGGESTED READINGS

- Stone, Carol Leth. The Basics of Biology. The United Kingdom, Greenwood Press, 2004.

Dynamic QR Code for Further Reading

2 Classification

UNIT SPECIFICS

Through this unit we have discussed the following aspects:

- *Cellular foundation of life*
- *Explained the biological classification system*
- *Described the classification based on cellularity, Energy utilization, Excretion, etc.*
- *Model organisms and their significance*

The ***practical applications of the topics are discussed*** *for generating further curiosity and creativity as well as improving problem solving capacity.*

Besides giving a large number ***of multiple choice questions as well as questions of short and long answer types*** *marked in two categories following lower and higher order of* ***Bloom's taxonomy****, assignments through a number of numerical problems, a list of references and suggested readings are given in the unit so that one can go through them for practice. It is important to note that for getting more information on various topics of interest some* ***QR codes*** *have been provided in different sections which can be scanned for relevant supportive knowledge.*

After the related practical, based on the content, there is a ***"Know More"*** *section. This section has been carefully designed so that the supplementary information provided in this part becomes beneficial for the users of the book. This section mainly highlights the initial activity, examples of some interesting facts, analogy, history of the development of the subject focusing the salient observations and finding, timelines starting from the development of the concerned topics up to the recent time, applications of the subject matter for our day-to-day real life or/and industrial applications on variety of aspects, case study related to environmental, sustainability, social and ethical issues whichever applicable, and finally inquisitiveness and curiosity topics of the unit.*

RATIONALE

To convey that classification per se is not what biology is all about. The underlying criterion, such as morphological, biochemical, or ecological,is highlighted—the hierarchy of life forms at the phenomenological level. A common thread weaves this hierarchy classification.

PRE-REQUISITES

Biology: Fundamental unit of life, Reproduction (Class XII)

UNIT OUTCOMES

List of outcomes of this unit is as follows:

U1-O1: Discussed the classification based on cellularity, energy, carbon utilization, and excretion.

U1-O2: Described the various habitats

U1-O3: Discussed the importance and characteristics of model organisms.

Unit-2 Outcomes	***EXPECTED MAPPING WITH COURSE OUTCOMES*** *(1- Weak Correlation; 2- Medium correlation; 3- Strong Correlation)*				
	CO-1	***CO-2***	***CO-3***	***CO-4***	***CO-5***
U2-O1	*3*	*2*	*2*	*1*	*1*
U2-O2	*3*	*3*	*1*	*1*	*2*
U2-O3	*3*	*2*	*2*	*1*	*1*

2.1 INTRODUCTION

Our planet is teeming with living organisms, each of which draws its supply of matter from its surroundings and uses these surroundings as a source of raw materials for the production of copies of themselves. Order, sensitivity, responsiveness to stimuli, adaptability, growth and development and control, chemotaxis, and energy processing are some of the essential traits and functions that are shared by all groups of living creatures. Every living thing is composed of cells, which all have the same basic make-up and components on the inside, and share the same machinery. On the surface of the earth, living creatures may be found in every conceivable setting, which gives life incredible diversity. Despite this, all forms of life share the traits and functions that have been discussed previously. Approximately 3.7 billion years ago, life began to develop and adapt, eventually leading to the formation of diverse species that give rise to offspring that are members of the same species. Even though the vast majority of living things are either single-celled or multicellular systems in which groups of cells perform specific activities and are linked by communication networks, the single cell is nonetheless the carrier for the hereditary information that identifies species.

2.2 HIERARCHY OF LIFE FORMS AT PHENOMENOLOGICAL LEVEL

Hierarchies are important for evolutionary biologists to understand how living systems have evolved and to theorize how extinct animals functioned based on their position on the hierarchy. Hierarchy is also a feature of gene-regulatory networks, which play an important role in organism development. It is basically a classification according to a hierarchy, or levels and orders. Classification is the process of grouping anything into useful categories (according to a systematic plan or order) based on some easily discernible characteristics. In this chapter, we will study different hierarchical classification systems that have evolved over a period of time.

2.2.1 Traditional classification

Living things are highly organized and structured, with a hierarchical structure that can be examined on a scale ranging from small to large. The atom is the most basic and smallest unit of matter. It is made up of a nucleus that is surrounded by electrons. Two or more atoms combine to form molecules, which are held together chemically by a chemical bond. Many biologically important molecules are macromolecules that are typically formed through polymerization (a polymer is a large molecule that is made by combining smaller units called monomers, which are simpler than macromolecules). A macromolecule is deoxyribonucleic acid (DNA), which contains the blueprints for the structure and operation of all living things.

Cells combine to form tissues in higher animals. Tissues are collections of similar cells that perform similar or related functions. Organs and organ systems are made up of these tissues. Organs exist in both animals and plants. Organ systems are higher levels of organization made up of functionally related organs. The circulatory system, for example, transports blood throughout the body and to and from the lungs; it includes organs like the heart and blood vessels. Furthermore, the classification includes population, which is defined as all of the individuals of a species living in a specific area. A community is a collection of all the different species that live in a specific area. A community then gives rise to an ecosystem, which is made up of all the abiotic and non-living elements of the environment, such as soil, rainwater, and so on, as well as living things. The biosphere, which represents the world's life zones, is the highest level in the hierarchy and includes all ecosystems, including air, water, and, to some extent, land. The biological levels of organization of living things are organelles, cells, tissues, organs, organ systems, organisms, populations, communities, ecosystems, and biospheres. As a result, the hierarchy of life forms is made up of all of these levels, which range from organelles to the biosphere.

2.2.2 Biological classification

Biological classification is a scientific method for classifying organisms into hierarchical groups and subgroups based on similarities and differences. The goal of biological classification is to group together all of the known plants and animals into categories that can be named, remembered, and studied. Various classification systems were developed to study these organisms and know their origin and evolution.

- Artificial system of classification focused mainly on similar/different morphological or vegetative characteristics for grouping or differentiating the organisms from one another.
- A natural system of classification is based to find similarities among organisms on basis of morphology, cytology (cell structure), phytochemistry (chemicals found in plants), embryology, and anatomy.
- The phylogenetic classification system is based on the phylogenetic trees to classify organisms based on their evolutionary descent and relationship.

According to biological classification history, Aristotle, a Greek philosopher, classified various animals based on habitat, characteristics, and so on. Later, during the 18th century, a Swedish botanist named Carolus Linnaeus introduced Taxonomic Hierarchy Categories, which is still used today. In this system, organisms were divided based on similar physical traits, such as the number of legs other and the shape of leaves. Since he had introduced this system, he was known as the "father of taxonomy." Taxonomy is a branch of biology that refers to the classification of various living species. A taxon is a group of organisms that are classified as a unit. Each level of the hierarchy is referred to as a taxonomic category or rank. Kingdom is always ranked first in this classification system, followed by division, class, order, family, genus, and species. The major taxonomic hierarchies are; Domain - highest level of classification and broadest category, Kingdom - The highest level of classification, the kingdom, is broken down into numerous degrees of subgroups, Phylum - The phylum is more specific than a kingdom, Classes: The organisms of a phylum are further sorted into more classes. It is one of the most common ranks put forth by Linnaeus, Order: One or more families that are similar to one another make up the order, Family: Here, the living organisms share some resemblance among themselves, Genus: It consists of several species with similar characteristics but different from that of species and another genus, Species: It describes a collection of organisms with comparable form, morphology, and reproductive characteristics. Sub-species can be found within a species.

2.2.3 Five kingdom classification system

During Linnaeus's time, a Two Kingdom classification system was introduced which divided Plantae and Animalia kingdoms that included all plants and animals, respectively. This system was in use until recently. However, this system made no distinction between eukaryotes and prokaryotes, unicellular and multicellular organisms, photosynthetic (green algae), and non-photosynthetic (fungi). As a result, the long-used two-kingdom classification was found to be inadequate. Hence, a need was also felt to classify organisms on basis of characteristics such as cell structure, wall nature, mode of nutrition, habitat, reproduction methods, evolutionary relationships, and so on.

As a result, R.H. Whittaker (1969) proposed a Five Kingdom Classification. The kingdoms defined by him were named Monera, Protista, Fungi, Plantae, and Animalia (as in Fig. 2.1). This approach aided in the differentiation of organisms based on their evolutionary relationships, mode of nourishment, mode of reproduction, and cell structure. Comparative information about the various traits of the five kingdoms is provided in Table 2.1. The kingdom Monera includes bacteria the most commonly occurring microorganism. E.g., Coccus, and Bacillus etc. The kingdom Protista includes **Chrysophytes (algae), Dinoflagellates, etc. The kingdom of Fungi** includes molds, mushrooms, yeast, etc. The kingdom Plantae includes all eukaryotes which have chloroplast. All multicellular eukaryotes, which are heterotrophs and lack cell walls, are placed under this kingdom.

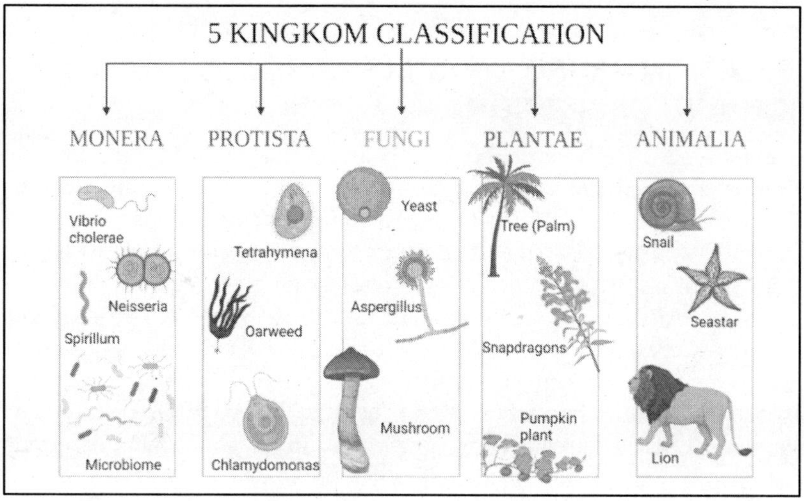

Fig. 2. 1 Classification system

(Images created using BioRender® software)

Table 2. 1

Characterization of five kingdom

	FIVE KINGDOMS				
	MONERA	PROTISTA	FUNGI	PLANTAE	ANIMALIA
Cell type	Prokaryotic	Eukaryotic	Eukaryotic	Eukaryotic	Eukaryotic
Cell wall	Noncellulosic (Polysaccharide + Amino acids)	Present in some	Present with chitin	Present (cellulose)	Absent
Nuclear membrane	Absent	Present	Present	Present	Present

Body organization	Cellular	Cellular	Multicellular /loose tissue	Tissue/organ	Tissue /organ system
Mode of nutrition	Autotrophic Chemosynthetic and Photosynthetic And Heterotrophic saprophytic/Parasitic	Autotrophic Photosynthetic and Heterotrophic	Heterotrophic saprophytic /parasitic	Autotrophic Photosynthetic	Heterotrophic/ Holozoic Saprophytic etc.
Genetic Recombination (if present)	Conjugation, transduction,and transformation	Fertilization and meiosis	Fertilization and meiosis	Fertilization and meiosis	Fertilization and meiosis

2.3 CLASSIFICATION BASED ON CELLULARITY, STRUCTURE, MODE OF NUTRITION AND EXCRETION

Biologists have classified organisms based on their similarities and differences. A biological classification is a systematic approach to organizing organisms into different groups and subgroups based on their similar and dissimilar traits. Aristotle, the ancient Greek philosopher who is frequently referred to as the "father of biology," is considered to be the originator of biological classification. He explained how to classify animals according to their habitat, which can be one of three things: air, water, or land. In the field of research concerning the animal kingdom, he was the first person to recognize the necessity of groups and names for groups. Later on, biologists began working on a classification system for living organisms based on the characteristics of the organisms themselves. There are many different routes one can take to explain characteristics. Certain attributes allow a group of organisms to be grouped together if they are sufficiently similar to one another. The manner in which something behaves and the roles that it plays are examples of its characteristics. Because of these characteristics, the classification of the organisms into groups can be accomplished. Some of the characteristics discussed in this section are;

- Cellularity: Unicellular or Multicellular
- Structure: Ultrastructure of Prokaryotes and Eukaryotes
- Energy and carbon utilization: Mode of nutrition
- Excretion: Different modes of eliminating wastes

2.3.1 Cellularity

Cells are the building blocks of all life forms. Depending on the composition, distribution, and number of cells present, organisms are classified as unicellular or multicellular (As in Fig. 2.2)

Unicellular organisms can be defined as the living-organisms, which consist of only a single cell. This single cell is capable to perform different life processes or cellular activities. It consists of prokaryotic organisms like bacteria and archaea, and eukaryotic organisms like protozoa, unicellular algae, and unicellular fungi. Most of the organisms commonly reproduce via asexual methods like fragmentation, budding, and binary fission, whereas few of them can also reproduce sexually via conjugation.

Multicellular organisms can be defined as living organisms, which consist of multiple cells. With these distinct cell organelles it is capable to perform different life processes or separate cellular activities inside a body. It only consists of eukaryotic organisms like insects, animals, birds, humans, etc. Most of the organisms commonly reproduce via sexual methods (by the formation of zygote), whereas few members few of them can also reproduce asexual means like budding, spore formation, etc.

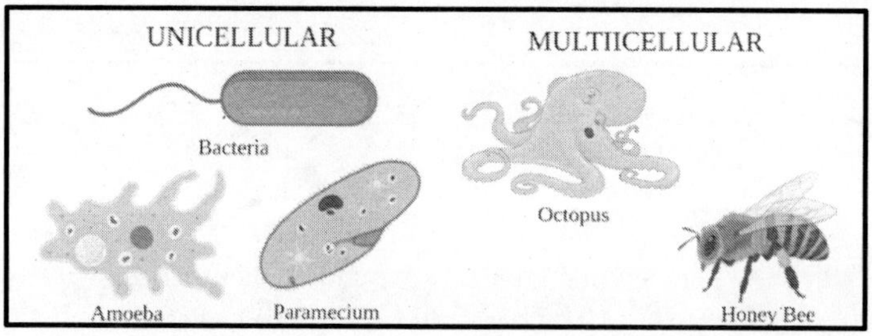

Fig. 2. 2 Structure of Unicellular and Multicellular organisms

(Images created using BioRender® software)

2.3.2 Ultrastructure of prokaryotes or eukaryotes

During the 1950s, all organisms were classified as prokaryotes and eukaryotes. The cells of prokaryotes are simple than eukaryotes. The basic features of cells of prokaryotes and eukaryotes are the plasma membrane and cytoplasm (As in Fig. 2.3). Eukaryotes cells form large and complex organisms. The characteristics of both types are mentioned below:

Prokaryotic cells are single-celled organisms that include bacteria and archaea. The nuclear membrane is absent and genetic material is present as a single chromosome. The organelles such as mitochondria, Golgi bodies, chloroplast, and lysosomes are absent. Flagella and pili are present for locomotion and attachment. The cell wall is composed of carbohydrates and amino acids and reproduction occur asexually by binary fission.

Eukaryotic cells are multi-celled organisms that include protozoa, fungi, plants, and animals. The nucleus is enclosed within the nuclear membrane and it contains single linear DNA which carries the genetic information. The organelle mitochondria is present. The locomotion is usually carried out with flagella and cilia.

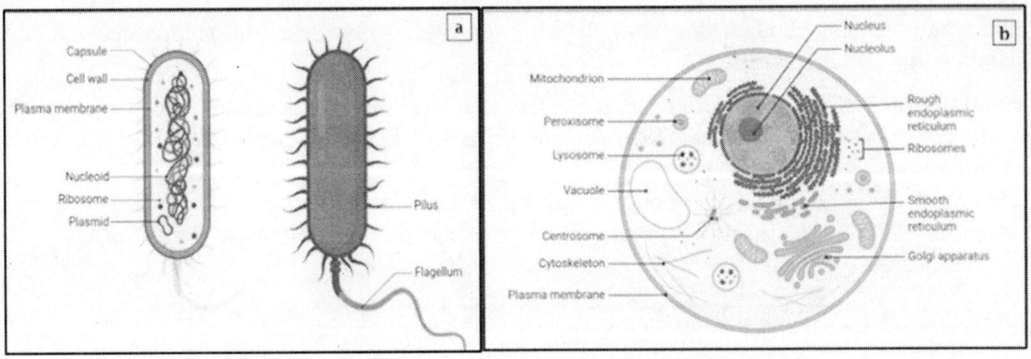

Fig. 2. 3 Ultrastructure of a) Prokaryote and b) Eukaryote

(Images created using BioRender® software)

2.3.3 Energy and carbon utilization

All living organisms, including plants and animals, require food. These organisms require food through various modes of nutrition in order to obtain energy as well as materials for growth and repair of damaged body parts. Food provides energy and nutrition for all living organisms on this planet. Nutrients are substances that provide energy and biomolecules required for various body functions. Nutrients are required by all living organisms for proper functioning and growth. However, they differ in how they meet this demand. To meet their nutritional needs, some animals consume simple inorganic compounds, while others consume complex compounds. The mode of nutrition differs between species.

Autotrophs

Organisms capable of producing their own food by utilizing the substance from the surrounding environment are called autotrophs. Energy generation can be carried out in either of two ways:

Photosynthesis: Photosynthesis occurs in the presence of light and chlorophyll. Water and carbon dioxide are broken down into glucose and oxygen. E.g. Green plants, Cyanobacteria, and green algae.

Chemosynthesis: Chemosynthesis utilizes the energy stored in inorganic compounds to produce carbohydrates, sulfur, and water. Eg. Sulfur bacteria, Iron-oxidizing bacteria, and nitrogen-fixing bacteria.

Heterotrophs

These organisms are unable to produce their own food and hence must obtain it from plants and animals. Two categories of heterotrophs exist;

Photoheterotrophs: The organisms can use sunlight as an energy source, but they are unable to use carbon dioxide as a source of carbon. Instead, the environment's substances including lipids, fatty acids, and carbs serve as carbon sources. For instance, heliobacteria, purple non-sulfur bacteria, and green non-sulfur bacteria.

Chemoheterotroph: The main supply of carbon for the organisms comes from organic compounds, which are also used as a source of energy. Eating other living or dead organisms provides energy. Animals, fungus, bacteria, and nearly all diseases, for instance.

Lithotrophs

These organisms use inorganic substrates like hydrogen sulfide, reduced iron compounds, etc., as a source of electron donors. Lithotrophs greatly aid the biogeochemical cycling of carbon, nitrogen, and sulphur. Nitrifying bacteria can convert NH_3 to NO_2 and NO_2 to NO_3; methanogenic archaea can remove electrons from H_2, add them to CO_2, and produce CH_4 (methane).

2.3.4 Excretion

The process of eliminating metabolic wastes (fluid, gaseous, organic, or inorganic) from the body is called excretion. Excess fluids, gases, and nutrients are eliminated through perspiration, urination, and defecation. These include water, gaseous wastes like carbon dioxide, and nitrogenous wastes like ammonia, urea, uric acid, and creatinine. An organism gets rid of these excess substances through the process of deamination. Toxic ammonia is either expelled in its original form or is first changed into a less harmful form, such as urea or uric acid, before being eliminated. Generally, three categories of animals can be distinguished depending on the nitrogenous wastes they create. The primary outcome of the deamination process is ammonia. However, it is very poisonous. Therefore, it needs to be diluted right away. Ammonia conversion becomes essential if there is no or restricted access to water. So, an organism's excretion mechanism is greatly influenced by water availability. Generally, there are three types:

Ammonotelism

The process involves the removal of nitrogenous wastes in the form of ammonia. Ammonia's basic nature disrupts the pH of the cell, making it unstable. Because ammonia is easily soluble in water, a large amount of water is required to reduce its toxicity. However, it is an energy-saving excretory mechanism. One gram of ammonia requires 300-500 ml of water to excrete. Aquatic invertebrates, bony fishes, and aquatic/larval amphibians all exhibit ammonotelism. Ammonotelic animals are those that lack an excretory system. E.g., Protozoa. Ammonotelic animals excrete ammonia through the skin, gills, and kidneys.

Ureotelism

The mode of excretion in which elimination of nitrogenous waste is in the form of urea is called as Ureotelism. Urea is comparatively less toxic to the body. Hence it can be stored in the body for a short period of time before it is excreted. It requires less amount of water for getting eliminated. The animals that follow this mode of excretion are known as ureotelic animals. Example - Humans, turtles, frogs, sharks etc.

Uricotelism

The mode of excretion in which elimination of nitrogenous waste is in the form of uric acid is called as uricotelism. The animals that follow this mode of excretion are known as uricotelic animals. Most of these animals live in dry regions or do not consume plenty of water (eg. birds), hence they must conserve water in their bodies. Uric acid is water insoluble crystals which requires very less amount of water to get eliminated from the body. Example - Birds (class Aves), *Helix* (commonly known as land snails), cockroach, lizard, snakes etc.

2.3.5 Habitat

A habitat is a natural environment in which a plant or animal lives (As in Fig. 2.4). Habitat meets all organism's environmental requirements for survival. A habitat's essential elements include shelter, water, food, and space. The following are examples of habitats:

Forests

Forest is a large area covered with plants. About one-third of our planet is covered in forests. Forests are home to 80% of the world's plant and animal species (many of which are yet to be discovered), and forests are home to 300 million people worldwide! Animal and plant species can be found in every layer of the forest, which explains why forest habitats are so diverse.Forest habitats are not only important for biodiversity, but they also significantly contribute to global oxygen levels. Trees absorb carbon dioxide and other greenhouse gases and emit oxygen in their place, aiding in the fight against climate change. Without forests and woodlands, the world would be less diverse in terms of plants and animals, and we would have far less oxygen to breathe.

<u>Tropical Evergreen Rainforest:</u> Only a small fraction of tropical forests can be classified as rainforests, which receive an annual rainfall of between 80 and 400 inches on average. This forest is distinguished by its extensive and dense vegetation, primarily made up of tall trees that reach various elevations.

<u>Tropical Deciduous Rainforest:</u> The predominant feature of a tropical deciduous rainforest is the presence of trees with broad leaves, in addition to dense understory vegetation such as bushes, shrubs, and the like. There are two primary seasons that are distinguishable in that location: summer and winter. This kind of forest can be found in a lot of different places around the world. This area is home to a very diverse collection of flora and fauna.

<u>Temperate Evergreen Forest:</u> A temperate evergreen forest is a type of forest that is distinguished from other types of forests by having a lower total tree count but a sufficient quantity of ferns and mosses.

<u>Temperate Deciduous Forest:</u>When there is an adequate amount of rainfall in a moist temperate region, a deciduous temperate forest will develop there. Also in this region, winter and summer are clearly distinguished from one another, and during the winter, trees lose their leaves. Tree species such as maple, oak, peach, and others predominate.

<u>Taiga/Boreal:</u> The Taiga is a forest region that lies immediately to the south of the Tundra and is dominated by evergreen conifers. Almost half of the year has an average temperature that is colder than the freezing point of water.

Deserts

Deserts are arid regions of land that are characterized by temperatures that are either extremely high or extremely low, along with low rainfall and little to no vegetation. Deserts are divided into two categories, hot deserts, and cold deserts, depending on the temperature conditions they experience.

<u>Sahara's hot and dry desert</u>

The Sahara Desert covers 8.54 million square kilometres. It's the world's largest, hottest desert. The Sahara Desert is 1,000 m above sea level. It includes Algeria, Tunisia, Egypt, Mali, Chad, Niger, Western Sahara, Sudan, Mauritania, Libya, and Morocco. The total area is 8,600,000 square kilometers, or 4,800 km east-to-west and 800 to 1,200 km north-to-south. This region's climate is hot, dry, and sizzling, with little annual precipitation. Desert days are sweltering. Daytime temperatures reach 45° C to 50° C, heating bare rocks and sand. Sometimes nights are below 0° C.

Ladakh's cold and desert

Ladakh is India's cold desert. It's found in the Great Himalayas in eastern Jammu and Kashmir and the western Himalayas in Himachal Pradesh, India. Ladakh's altitude ranges from 3,000 to 8,000 meters. High altitude causes a cold, dry climate. Summer days are just above 0° C and nights are below –30° C. Ladakh has several glaciers, including Gangotri, and many rivers. Ladakh's main river is the Indus. Ladakh has few types of grass and shrubs for animals to graze. In the valleys are poplars, willow groves, and Salix shrubs.

Grasslands

Grasslands are an intermediate stage in ecological succession, covering land where climate and soil don't allow tree growth. Nearly a quarter of the land is grassland. Climate and soil affect what plants grow here. Grasslands cover low-rainfall, poor-quality soil areas. Low rainfall prevents tree and shrub growth but supports monsoon grass growth. Tropical grasslands have grasses and drought-resistant thorny trees. Badgers, foxes, asses, zebras, and antelope graze on dairy and leather grasslands. Rodents, reptiles, and insects thrive in grasslands.

Mountains and Polar Habitat

The coldest ecosystems on the planet are the polar and mountainous regions. The Arctic and Antarctica are two examples of them. The lowest temperature ever measured was -89.6°C at Vostok station, Antarctica in 1983. Polar Region is home to creatures like the polar bear, penguin, seal, and walrus.

Aquatic Habitat

Freshwater habitats include bodies of water such as lakes, ponds, rivers, oceans, and streams; marine habitats include bodies of water such as oceans, intertidal zones, reefs, seabeds, and so on; and both freshwater and marine habitats are included in the definition of an aquatic ecosystem. The habitat of water-dependent living things, such as animals, plants, and microbes, is referred to as an aquatic ecosystem. Aquatic habitat is divided into marine and freshwater aquatic ecosystems. Both marine and freshwater ecosystems have subtypes.

Ocean ecosphere: Five major oceans are the Pacific, Atlantic, Indian, Arctic, and Southern. The Pacific Ocean is the largest and deepest of these five, while the Atlantic is second. The Southern Ocean has the most Krill. Turtles, crustaceans, plankton, corals, shellfish, blue whale, sharks, tube worms, reptiles, etc. live in the oceans.

Estuaries: It's usually where a sea and rivers meet, so the water is saline and diluted compared to the marine ecosystem. Estuaries stimulate primary production and trap plant nutrients, making them biologically productive. Tidal marshes, river mouths, and coastal bays are estuaries.

Coral reefs: They are called the Rain Forest of Oceans because they have so much aquatic flora and fauna. Coral reefs form an aquatic ecosystem. Calcium carbonate binds reef-forming coral polyps. Stony corals with clumped polyps dominate coral reefs.

Coastal Environment: When land meets water, coastal ecosystems form. These ecosystems have a unique structure, variety, and energy flow. Plants and algae dominate the coastal bottom. Fauna includes insects, snails, fish, crabs, shrimp, and lobsters.

Freshwater ecosphere: This aquatic ecosystem covers less than 1% of the earth's surface and includes wetlands, lentic, and lotic ecosystems.

Wetlands: These marshy areas are often flooded and harbor flora and fauna. Water lilies, marshes, swamps, Northern Pikes, dragonflies, Green Heron, etc. live in wetlands.

Lentic ecosystems: Floating and rooted plants, algae, and invertebrates live in ponds and lakes. Lentic ecosystems include lakes and ponds. In these habitats, algae, rooted and floating-leaf plants, and crustaceans live.

Lotic ecosystems: These aquatic ecosystems have fast-flowing, one-way water. They attract insects like beetles, mayflies, and stoneflies. River dolphins, beavers, otters, eel, minnow, and trout live there.

Fig. 2. 4 Habitat in different systems

(Source: pxhere.com – (Creative commons licenses)

2.3.6 Molecular taxonomy

Molecular taxonomy helps to identify and differentiate among specimens based on interspecific differences. This allows using molecular tools to establish the genetic link between individuals belonging to different taxonomic groups. It mainly focuses on data derived from hereditary material.

The objectives of molecular taxonomy are as follows.

- Reconstruction of the genetic relationship between various species.
- Study of evolution.
- To determine the time of divergence occurring between the various species.

Molecules used in molecular taxonomy are as follows.

- DNA is comprised of nucleotides which are single monomeric units that further consist of three elements, i.e., sugar, a nitrogenous base, and a phosphate group. Deoxyribose sugar is present in DNA, comprising a phosphate group on one side and a nitrogenous base on the other side.
- RNA is made up of nucleotides, which are made up of a nitrogenous base, a pentose sugar, and one or more phosphate groups. The presence of an H rather than an OH at the 2′ positions distinguish the structure from DNA.

Techniques such as DNA-DNA hybridization targeted sequencing, and whole-genome sequencing enables analysis of genotypic features that underpin all organisms similarities and differences. Genotypic and phenotypic variation together reveal traces of evolution and provide insight for historical relationship. Technological advancements drive progress in molecular taxonomy. That advancement is currently high-throughput, low-cost next-generation DNA sequencing. Advancements continue to reduce the cost, speed, and availability of molecular methods such as those described above.

The techniques involve in molecular taxonomy are;

DNA-DNA Hybridization: DNA-DNA hybridization is a molecular biology technique used in genomics to determine the degree of genetic similarity between pools of DNA sequences. It is commonly used to calculate the genetic distance between two organisms and has been widely applied in phylogeny and taxonomy.

Targeted sequencing: Targeted sequencing is a quick and low-cost method for detecting known and novel variants in specific sets of genes or genomic regions.

Whole-genome sequencing: All organisms (bacteria, plants, and mammals) have a unique genetic code, or genome, made up of nucleotide bases (A, T, C, and G). Sequencing is the process of determining the order of bases. Whole genome sequencing is a laboratory procedure that determines the order of bases in an organism's genome in a single process.

Immunological techniques: Immunological techniques include methods to study the immune system and generate or use immunological reagents. Immunological methods involve producing and using antibodies to detect proteins in biological samples. Eg. Elisa, Immunoblotting etc.

2.4 MODEL ORGANISMS

A model organism is a non-human species that is extensively studied to understand biological phenomena, with the expectation that discoveries in the model organism will provide insight into the workings of other organisms. Researchers worldwide use model organisms. These organisms share many human genes, are easily maintained in the lab, and have short generation times, making genetic manipulations easy to study. When human experimentation would be unfeasible or unethical, model organisms are used. This strategy is made possible by common descent and the conservation of metabolic and developmental pathways and genetic material over evolution.

The advantages of model organisms are:

- Short lifespan
- Small, easy maintenance
- Large offspring producers
- Well studied genome
- Easy genetic manipulations
- Only a few ethical issues

2.4.1 Escherichia coli

- Escherichia coli is a Gram-negative rod-shaped (bacillus) bacterium frequently used as a model organism (Fig. 2.5).
- Factors such as its ability to grow quickly on low-cost media and the availability of molecular tools for genetic manipulation make E.coli an ideal model organism for molecular genetics.
- Compared to the human genome, the E. coli genome is relatively small (4.5 to 5.5 Mbp) and simple (nearly 3 billion bp).
- The generation time can be as short as 20 minutes, and culture maintenance is as simple as meeting nutritional requirements.
- Current research areas of interest for E.coli include host for genetic elements and protein synthesis.

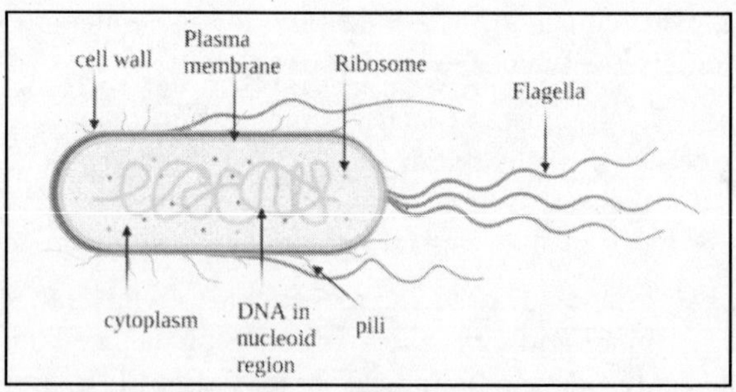

Fig. 2. 5 Structure of *Escherichia coli*

(Images created using BioRender® software)

2.4.2 *Saccharomyces cerevisiae*

- Saccharomyces cerevisiae is one of the most studied model organisms (Fig. 2.6)
- With a genome size of 12 Mbp, it is one of the simplest eukaryotes (containing membrane-bound organelles) and the first eukaryotic organism to be sequenced.
- Yeasts have simple nutritional needs and can be grown in standard laboratory conditions.
- The organism has many applications in the food industry and biotechnology.
- It is also used for cell division research and cancer research because it shares the same cell division process as humans.

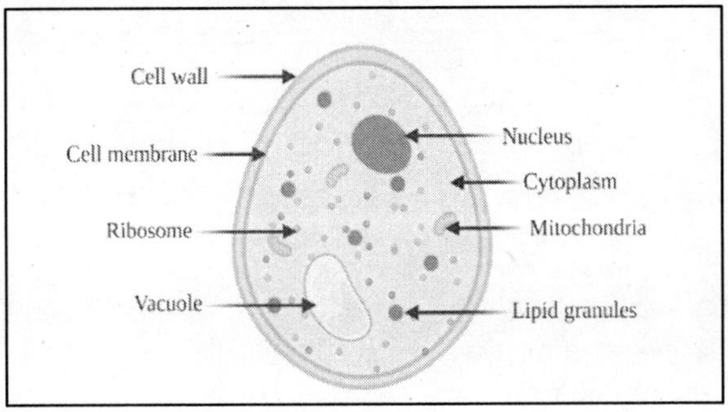

Fig. 2. 6 Structure of *Saccharomyces cerevisiae*

(Created with BioRender.com)

2.4.3 *Drosophila melanogaster*

Drosophila melanogaster has several characteristics that make it an ideal model organism (Fig. 2.7) including:

- Short life cycle: One of the main advantages is the short life cycle, which allows for large quantities of flies to be produced in a short period.
- Minimal culturing requirements: Because of their small size and minimal requirements, they are simple to maintain.
- Genetic manipulation: The presence of only four pairs of chromosomes in drosophila, compared to 23 pairs in humans, makes genetic transmission mapping easier. Drosophila's entire genome has been sequenced and annotated as the human genome. As a result, they were initially used in genetic studies.
- Drosophila has anatomical features (such as wings and eyes) that allow for easy characterization. Under a microscope, these genetic markers are easily identified.

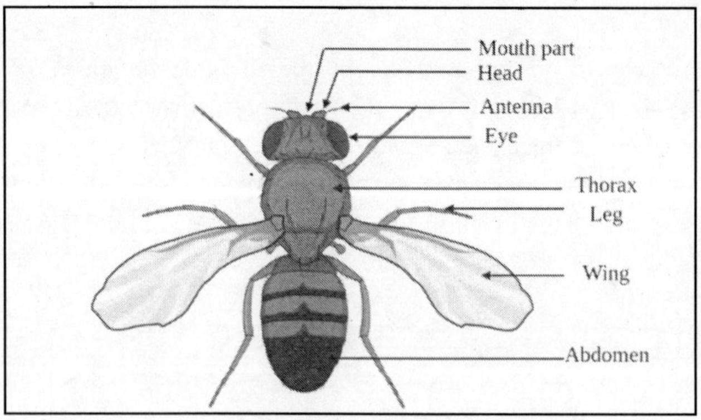

Fig. 2. 7 Structure of *Drosophila melanogaster*

(Images created using BioRender® software)

2.4.4 *Caenorhabditis elegans*

- *C. elegans* is a transparent, free-living nematode (Fig. 2.8).
- It is a multicellular eukaryotic organism that is an excellent model for studying the genetic control of development and physiology.
- *C. elegans* was the first multicellular organism to have its entire genome sequenced.
- Chemotaxis, thermotaxis, mechanotransduction, learning, memory, and mating behavior are all studied using the organism.

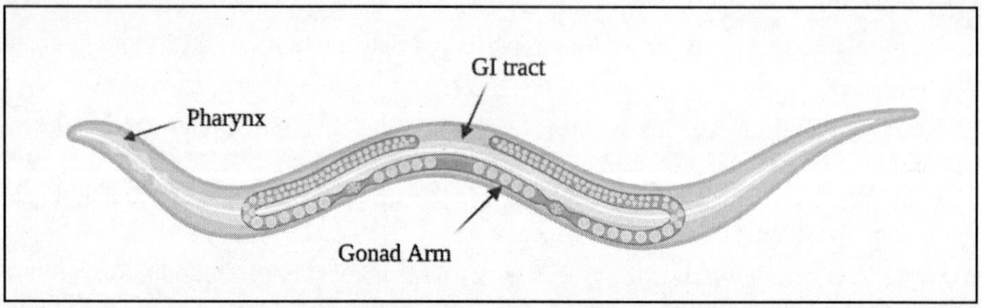

Fig. 2. 8 Structure of *C.elegans*

(Images created using BioRender® software)

2.4.5 *Arabidopsis thaliana*

- It is the most studied model organism in plant molecular genetics fundamental research (Fig. 2.9).
- Its small stature and genome, as well as its short generation time, allow for rapid genetic studies and molecular biology of many traits, such as flower development and light sensing.
- Arabidopsis was the first plant whose genome was sequenced.

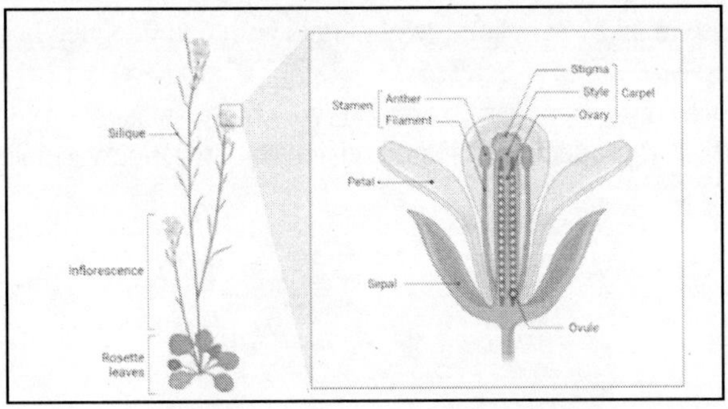

Fig. 2. 9 Structure of Arabidopsis thaliana

(Images created using BioRender® software)

2.4.6 Mus Musculus (mice)

- It is one of the simplest mammals to breed and keep in a laboratory (Fig. 2.10).
- It shares approximately 85% of its genes with humans, making it useful for studying human disease.
- It is used in developing vaccines for yellow fever and polio and the discovery of antibiotics such as penicillin and streptomycin.
- It is used for studies related to vaccine development, drug testing, toxicity studies, and immunological studies.
- It can be genetically modified to mimic Alzheimer's disease, anxiety and depression, cancer, cystic fibrosis, diabetes, obesity, and other diseases.

Fig. 2. 10 Mus musculus

(Source: pxhere.com – (Creative commons licenses)

2.4.7 Semnopithecus entellus (Monkey)

- Human biology and disease can be studied in Monkeys because they are very similar to humans; biologically, anatomically, and physiologically (Fig. 2.11).
- They have similar brains, muscle structures, and reproductive and immune systems. Therefore, research with monkeys gives us results that are more relevant to humans compared perhaps with the information obtained from mice or rats.

- However, because monkeys are complex animals with complex needs, they are only used when it is necessary.
- Monkeys are used in a range of research which includes neurological disorders such as Parkinson's disease, reproduction research, vision research, and vaccines.

Fig. 2. 11 Semnopithecus entellus

(Source: Wikimedia Commons-(Creative commons licenses)

In conclusion, we have discussed biological, cellularity, energy utilization, excretion, and habitat classification at the biological level. Besides, we have explained the role and importance of model organisms for research purposes.

UNIT SUMMARY

All living organisms are composed of basic building blocks that combine to form the fundamental unit of life known as cells. These simple biological units define all of their structural and functional abilities. An in-depth examination of the discovery of cells, including who discovered them and how, as well as their types, components, and functions, are carried out. Their classification is also discussed regarding cellularity, energy utilization, and habitat. In addition, we have expanded on the benefits of model organisms for research purposes.

EXERCISES

Multiple Choice Questions

1) Prokaryotes and eukaryotes generally have which of the following features in common?

   A. A membrane-bound nucleus
   B. A cell wall made up of cellulose
   C. Ribosomes
   D. Flagella or Cilia that contain microtubules

2) Prokaryotes are classified as belonging to two different domains. What are the domains?

   A. Bacteria and Eukarya

B. Archaea and Monera
C. Bacteria and Archaea
D. Bacteria and Protista

3) Which organisms do not obey cell theory?

A. Bacteria
B. Virus
C. Plants
D. Fungi

4) Which organelle of a eukaryotic cell is associated with protein maturation and transportation?

A. Vacuoles
B. Nucleus
C. Plasma membrane
D. Endoplasmic Reticulum

5) Taxonomic hierarchy refers to

A. Stepwise arrangement of all the categories for the classification of plants and animals
B. A group of senior taxonomists who decide the nomenclature of plants and animals
C. A list of botanists and zoologists who have worked on the taxonomy of a species or group
D. Classification of a species based on the fossil record

6) According to the five-kingdom system of classification, all unicellular eukaryotic organisms are included under Kingdom.

A. Protista
B. Monera
C. Fungi
D. Plantae

7) During photosynthesis, plants use light energy to convert water and CO_2 into

A. CO_2 and oxygen
B. Oxygen and water
C. Carbohydrates and water
D. Oxygen and carbohydrates

8) Organisms capable of using reduced inorganic compounds as electron donors are termed as

A. Lithotroph
B. Phototroph
C. Chemotroph
D. Photo-organotrophs

9) Organisms like purple non-sulfur bacteria, green-non sulfur bacteria, and heliobacteria are examples of

   A. Autotrophs
   B. Lithotrophs
   C. Chemoheterotrophs
   D. Photoheterotrophs

Answers: 1) C; 2) C; 3) B; 4) D; 5) A; 6) A; 7) D; 8) A

Short Answer Type Questions

1) Explain the components of Modern Cell Theory.
2) Describe the mode of nutrition of the Five Kingdoms.
3) Mention the levels of biological classification.
4) What are the carbon, energy, and electron sources of autotrophs, heterotrophs, and lithotrophs?
5) In short, describe the environment of the desert region.
6) Give reason: Why is *E. coli* considered a model organism?

Long Answer Type Questions

1) Explain the ultrastructure of prokaryotes and eukaryotes.
2) Define excretion. Explain the types of excretion.
3) Explain the characteristics of *D. melanogaster, C.elegans* and *A. thaliana* as model organisms.

KNOW MORE

- Some amazing and mysterious creatures, commonly known as flatworms, can reproduce asexually simply by tearing themselves in half. Each half then grows to become an individual worm.
- The human bladder can stretch to hold about 400ml of urine.
- All the blood in our body is filtered 400 times through the kidneys every day.
- There are more bacteria in a person's mouth than in the entire population of the world.

REFERENCES AND SUGGESTED READING

- Joanne, M. (2008). Prescott/Harley/Klein's Microbiology.
- Kathleen, P. T. (2017). *Foundations in Microbiology: basic principles*. MCGRAW-HILL US HIGHER ED.

- Madigan, M. T., Martinko, J. M., & Parker, J. (2006). *Brock biology of microorganisms* (Vol. 11, p. 136). Upper Saddle River, NJ: Pearson Prentice Hall.
- Salle, A. J. (1948). Fundamental principles of bacteriology. *Fundamental Principles of Bacteriology.*, (Edn 3).

Dynamic QR Code for Further Reading

3 Genetics

UNIT SPECIFICS

Through this unit we have discussed the following aspects:

- *Mendel's laws and Concept of allele*
- *Gene mapping and Gene interaction*
- *Mitosis and Meiosis*
- *Concept of recessive and dominance*
- *Concept of mapping of phenotype to genotype*
- *Single gene disorders in humans*
- *Concept of complementation using genetics*

The ***practical applications of the topics are discussed*** *for generating further curiosity and creativity as well as improving problem solving capacity.*

Besides giving a large number ***of multiple choice questions as well as questions of short and long answer types*** *marked in two categories following lower and higher order of* ***Bloom's taxonomy****, assignments through a several numerical problems, a list of references and suggested readings are given in the unit so that one can go through them for practice. It is important to note that for getting more information on various topics of interest some* ***QR codes*** *have been provided in different sections which can be scanned for relevant supportive knowledge.*

After the related practical, based on the content, there is a ***"Know More"*** *section. This section has been carefully designed so that the supplementary information provided in this part becomes beneficial for the users of the book. This section mainly highlights the initial activity, examples of some interesting facts, analogy, history of the development of the subject focusing the salient observations and finding, timelines starting from the development of the concerned topics up to the recent time, applications of the subject matter for our day-to-day real life or/and industrial* *applications on variety of aspects, case study related to environmental, sustainability, social and ethical issues whichever applicable, and finally inquisitiveness and curiosity topics of the unit.*

RATIONALE

To convey that "Genetics is to biology what Newton's laws are to Physical Sciences". Mendel's laws, Concept of segregation and independent assortment. Concept of the allele. Gene mapping, Gene interaction, Epistasis. Meiosis and Mitosis be taught as a part of genetics. Emphasis is to be given Not to the mechanics of cell division nor the phases but how genetic material passes from parent to offspring. Concepts of recessiveness and dominance. Concept of mapping of phenotype to genes. Discuss the single gene disorders in humans. Discuss the concept of complementation using human genetics.

PRE-REQUISITES

Biology: (Class XI and XII)

UNIT OUTCOMES

List of outcomes of this unit is as follows:

U3-O1: Gives overview about Genetics, Heredity, Mendel's law and Concept of allele

U3-O2: Explain gene mapping, gene interaction and Epistasis

U3-O3: Describe Meiosis and Mitosis from genetics point of view

U3-O4: Understanding chromosomal abnormalities and single gene disorder

U3-O5: Defining the concept of complementation using human genetics

Unit-3 Outcomes	EXPECTED MAPPING WITH COURSE OUTCOMES (1- Weak Correlation; 2- Medium correlation; 3- Strong Correlation)				
	CO-1	CO-2	CO-3	CO-4	CO-5
U3-O1	-	2	3	-	-
U3-O2	-	1	3	-	-
U3-O3	-	2	3	-	-
U3-O4	-	-	3	-	-
U3-O5	-	-	3	-	-

3.1 INTRODUCTION TO GENETICS

Genetics is the branch of biology that deals with the study of genetic variation, genes, and heredity. Genetics and its concepts had been observed for centuries, however, it was scientifically studied by Gregor Mendel. Genetics is the study of inheritance on several levels of comprehension, from molecules to populations. Genetics is central to contemporary biology, making its comprehension crucial for all life sciences scholars. Numerous areas of daily living are profoundly influenced by the field. The food we consume and the clothing we wear are derived from species that have been genetically modified. On the basis of fundamental genetic discoveries, the causes of significant human diseases are being uncovered, and treatments are being devised. The management of human health is increasingly dependent on genetic and genomic information. Genetics is a growth field since these effects will likely increase over the future decades.

3.1.1 Past, modern and future genetics

The Importance and role of genetics: We all have genes that have an effect on our life. They have an impact on our height, weight, hair colour, and skin pigmentation. They affect our susceptibility to several diseases and ailments and even contribute to our intelligence and character. Genes are vital to our being and identity. Although genetics is a relatively recent discipline, individuals have understood the heritable nature of traits and "practised" genetics for millennia. When humans began to apply genetic principles to the domestication of plants and animals, agriculture began to flourish. Today, the principal crops and animals used in agriculture have undergone considerable genetic modifications to significantly enhance their yields and give numerous desirable properties, including resistance to disease and pests, specific nutritional attributes, and harvest-facilitating characteristics. Today, a large amount of the food produced worldwide is derived from genetically modified corn, soybeans, and other crops. Numerous medications and food additives are also produced in the pharmaceutical business by fungus and bacteria that have been genetically modified to be effective makers of these chemicals. Commercially generated growth hormones, insulin, and clotting factor are now created by genetically modified bacteria. Additionally, genetics plays an important part in medicine. Physicians acknowledge numerous diagnostic tests have been made possible by the emergence of fundamental insights into molecular genetics. Without a good understanding of genes and genetic approaches, the study of practically any discipline of biology or medicine is insufficient.

Evolution of Genetic Variation: On Earth, a vast variety of life forms and characteristics inhabit virtually every imaginable area. All life has a common ancestor thus this diversity has developed during the past four billion years. Adaptation is another defining characteristic of life; many species are exquisitely adapted to their surroundings. The history of life is a chronicle of the emergence of new forms of life, the extinction of old forms, and the transformation of existing forms. Diversification and adaption of life are the results of evolution, which is merely genetic change over time. Evolution is a two-step process: first, genetic variants develop at random, and then the frequency of certain variants grows or decreases. Therefore, genetic variety is the basis of all evolutionary change and, eventually, the basis of all known life. Understanding the past, present, and future of life is dependent on genetics, the study of genetic diversity.

Divisions of Genetics: Genetics has traditionally been split into three key subfields: transmission genetics, molecular genetics, and population genetics. **Transmission genetics** involves the fundamental principles of genetics and how features are transmitted from one generation to the next. This field examines the connection between chromosomes and heredity, as well as the arrangement of genes on chromosomes and gene mapping. **Molecular genetics** is concerned with the chemical composition of the gene itself, including how genetic information is encoded, duplicated, and expressed. It consists of the cellular processes of replication, transcription, and translation which transfer genetic information from one molecule to another and gene regulation the mechanisms that control the expression of genetic information. **Population genetics** investigates the genetic makeup of groups of individuals belonging to the same species (populations) and how that genetic makeup varies over time and space. Population genetics is essentially the study of evolution because evolution involves genetic change. Population genetics focuses on the collection of genes present in a population.

3.1.2 Heredity and principles of heredity

As the twentieth century has unfolded, the new science of genetics has come to occupy an increasingly important position at the centre of the science of life. It is claimed that the origin of the science of genetics can be traced to one man - Gregor Mendel (1822-1884), raised in German-speaking Silesia, who entered the Augustinian order in the Monastery of Brünn, Moravia, and taught high school science, also finding time to conduct experiments in the hybridization of plants, before becoming abbot of his monastery. Mendel's research with pea plants is what made him most famous.

Fig. 3. 1 Gregor Johann Mendel 'Father of Genetics'

(Source: 5.10 Mendel's Experiments and Laws of Inheritance – Human Biology (*under creative commons licenses)*

On numerous occasions, it has been noted that family members have certain similar qualities, such as facial features, skin tones, etc. Why is this so? Why does a child resemble their mother in certain aspects and their father in others? In qualities that run-in families, hereditary variables, such as the genetic material a person inherits from his or her parents, have a role. All animals and plants share the same traits. Heredity, or the transmission of character qualities from one generation to the next, is the phenomenon of offspring inheriting their parents' characteristics. On the chromosomes are the genes responsible for the inheritance of characteristics. In addition, it has been observed that although while offspring receive qualities from their parents, they are unique and possess traits that distinguish them from their parents. Variations relate to these differences between progeny and parents. Genetics is the study of heredity and genetic variation from a scientific perspective.

The primary objective of biotechnology is to modify living organisms or alter their genetic composition in order to develop products that improve the quality of life for humans. Understanding genetics and the inheritance of traits is essential for using biotechnological techniques to change genes. To alter a trait, it is necessary to identify the genetic components (genes and their allelic forms at the population level) that influence the trait. The principles will be covered in this chapter.

3.2 MENDEL'S LAWS OF INHERITANCE

In the middle of the 19th century, advancements were made in understanding inheritance. After seven years of hybridization study on garden peas, Gregor Mendel (Fig. 3.1) formulated the rules of heredity in living creatures (1856–1863). Mendel's research on inheritance patterns initiated the application of statistical analysis and mathematical reasoning to biological problems. Greater accuracy in the data he gathered. In addition, the validation of his beliefs by research on successive generations of his test plants proved that his findings were theoretical conjectures and general rules of inheritance.

Mendel studied garden pea plant characteristics that were displayed as two opposite phenotypes, such as tall or dwarf plants or yellow or green seeds. As a result, he was able to formulate a fundamental set of inheritance rules that other scientists used to explain the complexity and diversity of natural data. Mendel tested artificial pollination and cross-pollination on a variety of pure-bred pea lines. A true-breeding line is one that exhibits steady trait inheritance and expression over numerous generations after having experienced continual self-pollination. Mendel chose fourteen true-breeding pea plant varieties, pairing them in ways that were identical but for one feature that was different. Smooth or wrinkled seeds, yellow or green seeds, inflated (full) or constricted green or yellow pods, tall or dwarf plants, violet or white blossom colour, and axial or terminal flower location were some of the contrasting features chosen (Fig. 3.2).

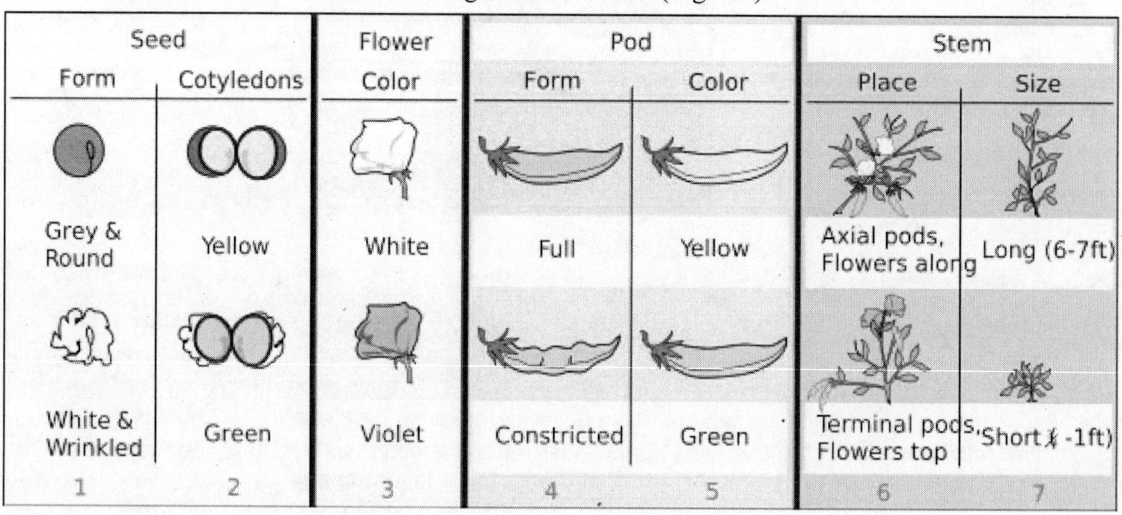

Fig. 3. 2 Contrasting Traits Studied by Mendel in Pea plant

(Source: https://humanbiology.pressbooks.tru.ca/chapter/5-9-mendels-experiments-and-laws-of-inheritance/) *under Creative Commons licenses*

Why did Mendel choose the Pea plant for his experiments?

1. The pea plant is simple to grow and maintain.
2. They are naturally self-pollinating, but also capable of cross-pollination.
3. Due to the fact that it is an annual plant, multiple generations can be studied in a short amount of time.
4. It contains contrasting characters

3.3 LAWS OF INHERITANCE PROPOSED BY MENDEL:

Mendel structured his experiments in a way that he would observe one pair of contrasting characters at one time. He began his experiments using purebred lines for contrasting characters. He cross-pollinated two pure lines for contrasting characters and the resultant offspring were called the F1 generation (also called the first filial generation/ monohybrid cross). The F1 generations were then self-pollinated which gave rise to the F2 generation the of second filial generation or dihybrid cross. These two experiments lead to the formulation of Mendel's laws known as laws of inheritance which are:

- **Law of Dominance**
- **Law of Independent Assortment**
- **Law of Segregation**

3.3.1 Law of Dominance

The first law of inheritance is known as Mendel's law. Only the dominant trait in the phenotypic will be passed down to hybrid offspring, in accordance with the law of dominance. Recessive characteristics are those alleles that are suppressed, whereas dominant traits are those alleles that control the trait. (Fig. 3.3)

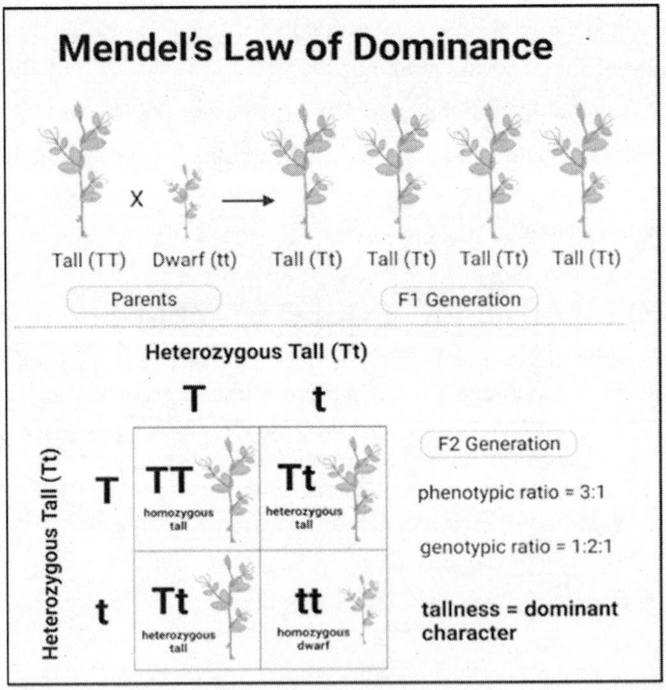

Fig. 3. 3 Law of Dominance

(Source: https://thebiologynotes.com/mendels-law-of-dominance/)

"When parents with pure, contrasting traits are crossed together, only one form of trait appears in the next generation. The hybrid offspring will exhibit only the dominant trait in the phenotype."

According to the law of dominance, only one of the contrasting features of the parents will be displayed in the F1 generation, and both parents' traits will be expressed in the F2 generation in a 3:1 ratio. A recessive trait is one that is suppressed while a dominant trait is one that is displayed in the F1 generation. The law of dominance essentially states that the dominant trait always dominated or masked the recessive traits. The Mendel experiment can be used to explain this law.

A monohybrid cross is created when two monohybrid traits are combined (TT and tt). Here, identical plants that only differed by one character were crossed. Mendel started with a pair of pea plants that had two distinct features, one tall and the other dwarf, for the monohybrid cross. Cross-pollination between tall and dwarf plants produced tall plants, known as F1 progeny. Dominant traits are those that are manifested in the phenotype, whereas recessive traits are those that are not. He then carried out further studies on the self-pollination of F1 offspring plants. Due to the 3:1 ratio of tall to short plants produced as a result, the law of segregation was formed.

3.3.2 Law of Independent Assortment

Also known as ***Mendel's second law of inheritance, it states that a pair of traits segregate independently of another pair during gamete formation.*** As the individual heredity factors assort independently, different traits get equal opportunities to occur together.

Let us now consider a dihybrid cross between homozygous round shape and yellow colour (RRYY) seeded pea plant with a homozygous wrinkled and green colour (rryy) seeded pea plant. All F1 progeny were round seeded having yellow colour. In this following example which traits are dominant, and which are recessive? In F1 progeny, as all plants were round and yellow seeded, it clearly showed that they are dominant over wrinkled and green seeded traits. The result of F2 generation upon selfing is explained in Fig. 3 in which a ratio of 9:3:3:1 of offspring with 9 round yellow, 3 wrinkled yellow, 3 round green, and 1 wrinkled green (9:3:3:1) is observed. Since two pairs of contrasting characters are included in such crosses, hence they are called dihybrid crosses. Based upon such observations on dihybrid crosses, the third principle of inheritance, i.e., **Law of Independent Assortment** was proposed. (Fig. 3.4)

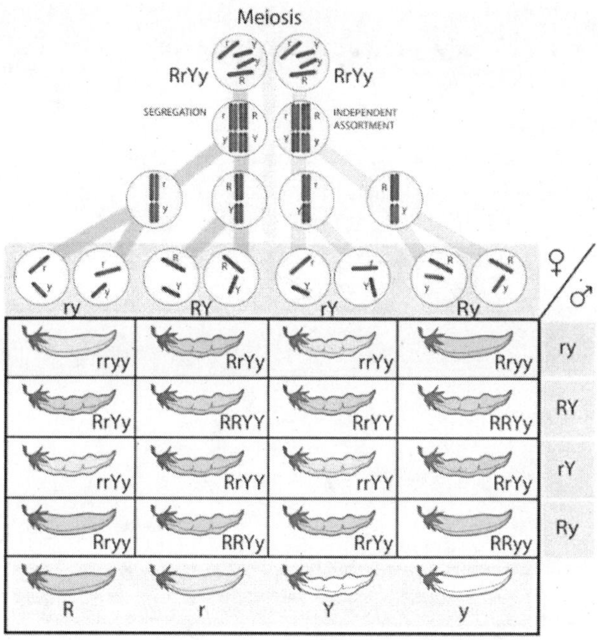

Fig. 3. 4 Law of Independent Assortment

(Source: https://biologydictionary.net/law-of-independent-assortment/) *under Creative Commons* licenses

3.3.3 Law of Segregation

The law of segregation states that during the production of gametes, two copies of each hereditary factor segregate so that offspring acquire one factor from each parent. In other words, allele (alternative form of the gene) pairs segregate during the formation of gamete and re-unite randomly during fertilization. This is also known as Mendel's third law of inheritance.

When Mendel cross pollinated a pure (homozygous) tall pea plant with a pure dwarf pea plant, he noticed that the progeny of first generation (First filia or F1 generation, which was raised by collecting the seeds produced from this cross) were all tall. The dwarf phenotype was missing. What happened to the dwarf trait? When the said F1 offspring were self-pollinated to raise F2 generation, surprisingly both tall and dwarf plants appeared in the ratio of 3:1 (3 tall and dwarf). Since Mendel designed this experiment by considering only one contrasting trait, i.e., tall and dwarf, this cross is called monohybrid cross. Interestingly, in all such monohybrid crosses involving other contrasting pair of characters carried out by Mendel, similar ratio of approximately 3:1 was obtained in F2 generation. These results prompted Mendel to propose that each individual has two factors for each character (trait) and that one factor (which was later named as gene) was inherited from each parent through gametes.

This is the reason that the dwarf feature which was not there in F1 generation was found in F2. Hence, F1 tall plants are heterozygotes as they contain two different alleles (Tt). As F1 plants are heterozygous tall (Tt), this indicates that the tall allele (T) is dominant over dwarf allele (t). Thus, dwarf allele (t) is recessive to tall allele (T). Understanding of these crosses can be well understood by the graphical representation developed by Reginald C. Punnett, a British geneticist. Using Punnett Square, we can easily calculate the probability of all possible genetic combinations or genotypes.

We can see in (Fig. 3.5), that when plants in F1 heterozygous progeny were self-pollinated as they produced 'T' and 't' gametes, the progeny revealed three genotype combinations; TT, Tt, tt in a ratio of 1:2:1 respectively. Here we learnt that through Punnett Square by using mathematics, we can easily calculate the probability of genotype (genetic make-up) and phenotype (morphological or observable traits) of future progeny. This clearly shows that the phenotypic ratio of a monohybrid cross is 3:1 and the genotypic ratio is 1:2:1.

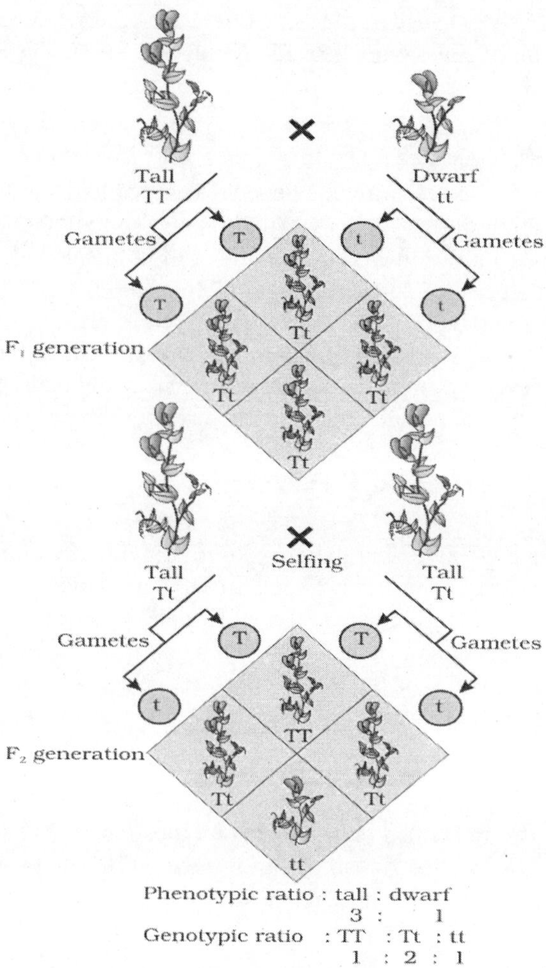

Fig. 3. 5 Segregation of height character in pea plant

(Source: https://www.pmfias.com/inheritance-mendels-laws-of-inheritance-mendels-experiments-on-inheritance/)

3.4 CONCEPT OF ALLELE

Each gene has two alternative versions known as alleles. When two identical alleles for a characteristic are present in an individual, they are called homozygous alleles. When two different alleles for a trait are present, they are called heterozygous alleles.

Any one of two or more genes that may alternately appear at a specific location (locus) on a chromosome is known as an allele, also known as an allelomorph. Alleles may exist in pairs or there may be several alleles influencing how a certain trait is expressed (phenotype). The genotype of an organism is made up of the alleles it possesses in combination. The genotype of an organism is referred to as homozygous or heterozygous depending on whether the paired alleles are the same or different. In a heterozygous pairing, a dominant allele will take precedence over a recessive allele's features. However, some alleles may be co-dominant, which means that neither one operates as dominant or recessive. The human ABO blood group system serves as an illustration; people with type AB blood have one allele for A and one for B. Type O people fall into this category.

The majority of characteristics are determined by more than two alleles for a particular gene encoding a characteristic. Multiple variants of the allele may occur; nevertheless, a diploid individual can only carry two alleles of a gene. During the generation of haploid gametes, only one will bind to the assigned gene location during meiosis. Additionally, some features are governed by many gene locations. Both possibilities increase the number of implicated alleles. All genetic qualities result from interactions between alleles. Mutation, crossing over, and environmental variables alter the frequency of phenotypes (and consequently their alleles) in a population. For instance, alleles carried by individuals with high fitness (meaning they successfully reproduce and pass on their genes to their kids) are more likely to endure in a population than alleles carried by individuals with lower fitness, which are gradually lost over time.

3.5 GENE MAPPING

As the distance between 2 genes increases, the crossing over will be more and hence more will be recombinants. The greater the distance between linked genes greater the chance of crossing over and the greater will be the number of recombinants. The unit of gene distance is cM (CentiMorgan) or mu (Map units). If the recombination frequency is 1% = 1cM distance between two genes on the chromosome. The highest possible frequency is 50% (In the case of Mendelism) which means the genes are either on 2 extreme ends of chromosomes or on different chromosomes.

For Example,

Suppose, two genes (A and B) are located on the same chromosome. An AA BB individual is crossed to an aa bb individual to produce Aa Bb offspring. The AaBb offspring are then test-crossed to aa bb individuals. Let us assume this produces a total of 400 offspring. Among these 325 offspring are parental and 75 are recombinant.

F2 Progeny: ABab = 160, abab = 165, Abab = 36 and aBab = 39

With the help of the above data, we can calculate the recombination frequency and map distance in the following way:

= (36+39/36+39+160+165) x 100

= 18.8 cm

$$\text{Recombination Frequency} = \frac{\text{Total Number of Recombinants}}{\text{Total Number of Offspring}} \times 100$$

The genes are approximately 18.8 cm apart.

3.6GENE INTERACTION

Genetic interaction is the set of functional associations between genes. One such relationship is epistasis, which is the interaction of non-allelic genes in which the effect of one gene is masked by another gene, either causing the effect to be suppressed or resulting in the production of a new phenotype.

3.6.1 Epistasis

Epistasis is a circumstance where the expression of one gene is modified (e.g., masked, inhibited or suppressed) by the expression of one or more other genes. Genes can either work together to create a new characteristic or mask one another so that one is recognized as "dominant." The phenotype of specific traits is determined by the conditional linkage between two genes. At each location, there are two alleles that influence phenotypes. They may interact in a way that makes one gene recessive to a dominant allele of the other, regardless of the genotype of the other gene. Tables and ratio charts are two other ways to express epistasis. There are four alleles for each of the two genes, creating a total of 16 potential pairs. There are sixteen phenotypes corresponding to these sixteen allele combinations. Not all combinations are unique because of the dominant and recessive traits of the dominant and recessive alleles. To visually represent the 16 potential allele pairings for four alleles, a 44 chart can be utilised. The colour of many bee species is demonstrated in the table below. (Fig. 3.6)

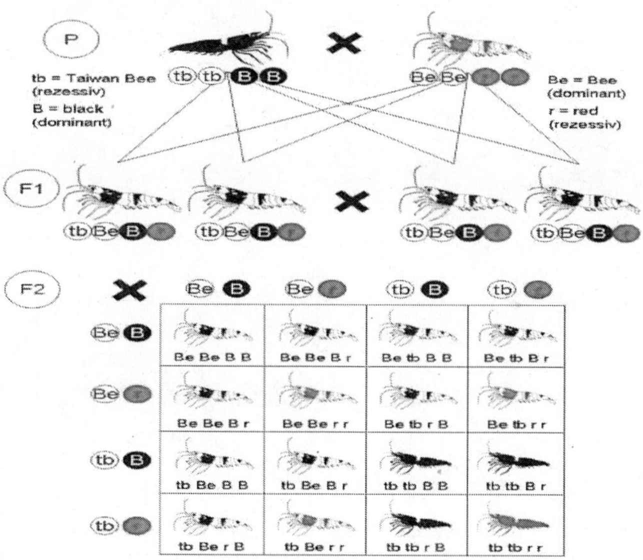

Fig. 3. 6 Example of Epistasis in Bee

(Source: https://biologydictionary.net/epistasis/)

3.6.2 Types of Epistasis

There are six common types of epistasis gene interactions: dominant, dominant inhibitory, duplicate dominant, duplicate recessive, polymeric gene interaction, and recessive.

Dominant epistasis, sometimes referred to as simple epistasis, occurs when a dominant allele suppresses the expression of both dominant and recessive alleles at a distinct location. Recessive epistasis is the process through which an expression is hidden by a recessive allele. Some genes have the ability to stop other genes from being expressed. Due to the gene's role as a suppressor or factor that prevents the expression of another allele, this phenomenon is known as dominant inhibitory or suppression epistasis.

Duplicate types of epistasis depend on two loci. Duplicate dominant epistasis, also referred to as duplicate gene action, occurs when a dominant allele conceals the expression of recessive alleles at two loci. Duplicate recessive epistasis occurs when a recessive allele conceals the expression of dominant alleles at two loci. Because both genes are necessary for the right phenotype to exist, it is often referred to as complementary gene action. The combination of two dominant alleles that enhances the phenotype or generates a median variation is known as polymeric gene interaction. Each dominant allele alone results in a physical characteristic that is distinct from the combined dominant alleles. This results in three phenotypes being produced from just two dominant alleles. It is clear from this that neither dominant allele is prevailing over the other.

Summer squash can have three different colours: white, yellow, and green. The dominant gene controls the colour white, the dominant gene G controls the colour yellow, and the recessive genes w and g control the colour green. In contrast to yellow and green, white is the prevailing colour. Epistatic refers to the relationship between the dominant and recessive G/g alleles and the dominant white allele. Following that, this interaction is categorized as simple or dominant epistasis.

3.7 MITOSIS AND MEOSIS

Handmade graphic depictions of mitotic chromosomes by Walther Flemming and meiotic chromosomes by Walter Sutton provided an early record of the physical path of chromosomes during cell division. The physical movement of chromosomes could then be correlated with cells' patterns of genetic inheritance. (The idea that genes were carried on cytological structures is now known as the chromosome theory) Using such methods, researchers determined that although mitosis and meiosis are both forms of cell division, the results of these processes are actually quite different.

3.7.1 Gene Transmission in Mitosis

Somatic cells undergo mitosis, which means that all cell types whose function is not the generation of gametes go through this process. Each chromosome is duplicated before each mitotic division, resulting in a full set of chromosomes in the nucleus of each new cell after division. Indeed, due to the inheritance of the same chromosome set and the same biological milieu, each succeeding duplicate cell will have the same genetic makeup as its parent, barring random mutations. This works well for repairing damaged tissue as well as for embryonic growth and expansion. (Fig. 3.7)

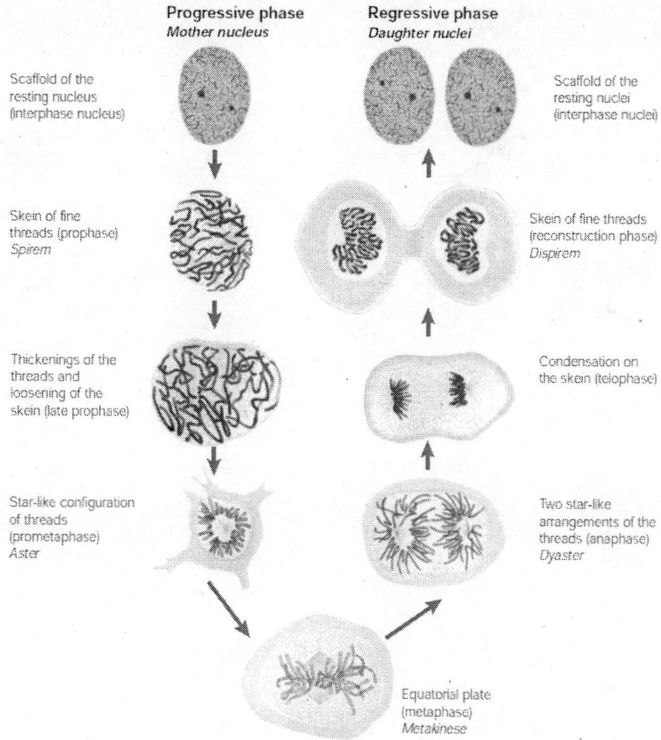

Fig. 3. 7 The progressive and regressive phases of cell division

(Source: https://www.nature.com/scitable/topicpage/mitosis-meiosis-and-inheritance-476/#)

All mitotic progenies are genetically similar because the genes found in the duplicate chromosomes are passed down to each succeeding cellular generation. There are, however, certain exceptions. For instance, spontaneous mutations that occur during mitotic division can lead to genetic differences in clonal species like bacteria. Furthermore, chromosomes can replicate several times without a cell division taking place in between. For instance, this happens in the cells of the salivary glands of Drosophila larvae, where there is a high metabolic requirement. When compared to the chromosomes in other Drosophila cells, these chromosomes, known as polytene chromosomes, are enormous. Without any cytokinesis, these chromosomes replicate by going through the first stages of mitosis, where the same cell includes dense configurations of duplicate chromosomes placed side by side that resemble thick rope-like strands. These chromosomes are thought to be hyper-replicated in order to facilitate the quick and abundant production of certain proteins that support larval growth and metamorphosis.

3.7.2 Gene Transmission in Meiosis

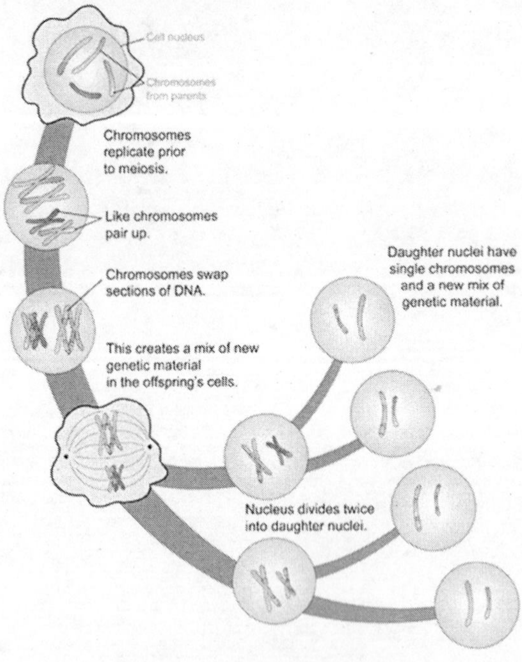

Fig. 3. 8 Different phases involved in Meiosis

(Source: https://www.ck12.org/book/ck-12-biology/section/5.3/)

Walther Flemming witnessed spermatozoa go through meiosis in 1882, but he thought it was mitosis. Flemming did note, however, that during spermatozoan production, chromosomes occur in pairs, in contrast to normal cell division. This finding, together with Sutton's laborious chromosome counting throughout the development of grasshopper sperm cells in 1902, proved beyond a doubt that cell division in gametes involved more than just mitosis. Sutton proved that the number of chromosomes was reduced during reductive division, also known as the division of spermatozoan cells. Sutton found that as a result of this process, each gamete contained half the genetic information of the original cell. A short while later, J. Farmer, B., and J. Meiosis, often known as this process, is the primary mechanism by which plants and animals produce gametes, according to E. S. Moore (Farmer & Moore, 1905).

More than anything else, Sutton's study has had the biggest impact by supporting Mendel's principle of independent assortment. Sutton observed that there was never a consistent maternal or paternal side to the cell division and that the location of each chromosome at the midline during metaphase was unpredictable. Each chromosome was therefore independent of the others. As a result, when the parent cell divided into gametes, each daughter cell's set of chromosomes may have included a combination of the parental features, however, this combination may not have been the same as in other daughter cells. (Fig. 3.8). To demonstrate this idea, think about the variation resulting from just three fictitious chromosomal pairs, as shown in the example below (Hirsch, 1963). One maternal and one paternal homologue make up each pair. The maternal chromosome is denoted by capital letters, whereas the paternal chromosome is denoted by lowercase letters in this sentence:

- Pair 1: A and a
- Pair 2: B and b
- Pair 3: C and c

When these chromosome pairs are reshuffled through the independent assortment, they can produce eight possible combinations in the resulting gametes:

A B C, A B c, A b c, A b C, a B C, a B c, a b C, a b c

3.8 CONCEPT OF MAPPING PHENOTYPE TO GENOTYPE

The relationship between the genotype, the genetic instructions encoded into a genome, and phenotype, the macroscopic realization of such instructions, remains mostly uncharted. In addition, tools able to uncover the connection between the phenotype with a specific set of responsible genes are still under definition. In one of the studies related to this, the focus was on yeast organelles called vacuoles, which are cell membrane compartments that vary in size and shape in response to various stimuli, and we develop a framework relating changes in cellular morphology to genetic modification. The approach combines a convolutional neural network (CNN) with an unsupervised learning pipeline and a segmentation, classification, and anomaly detection algorithm that are all based on deep learning.

The shape and structure of cellular organelles, such as the nucleus, mitochondria, and vacuoles, which are compartments designed specifically for carrying out biochemical operations, are related to the biological processes they mediate. Therefore, being able to control and influence the outcome of specific biochemical events would mean being able to design a specific organelle morphology. Significant new theoretical paradigms for the computational design of cellular structures have been developed through recent developments in computers. It's not very novel to connect cellular shape in model organisms like yeast to mutations. However, because of the inherent biological heterogeneity, it is clear that creating a sizable and thorough training set whether annotated or not is a significant bottleneck. We provide a hybrid supervised-unsupervised learning strategy that tries to accurately execute end-to-end mapping between existing morphologies and genetic damage in budding yeast organelles in order to address this problem.

Methodology, which is composed of several steps can be summarized as follows: after the cellular images are acquired, a segmentation step is performed using a U-Net CNN trained on a small fraction of the dataset. The segmented images are then analysed and a set of features is extracted from the masks. Images are partitioned in an unsupervised fashion with the number of classes automatically inferred from the data, and the resulting clusters are used as the training set for another CNN, which is ultimately used to classify the cells. Our results suggest that this methodology can reveal the relationship between phenotype and genotype in an accurate and unbiased way.

Even though the mechanistic design of cells and cellular structures is still a long-term research objective, developing tools that may infer biological design principles is a crucial first step towards the mechanistic realisation of biological systems for practical uses. An organelle that can function as a biological reactor for the synthesis of chemicals is the yeast vacuole. We can link various vacuole shapes to genetic disturbances using our system. In order to accomplish this, we use a mixed supervised-unsupervised learning methodology in an effort to lessen the load of annotation and the inherent bias associated with the human annotation work. Our findings demonstrate that the genotype-cellular phenotypic relationship may be determined with high accuracy and little user involvement. We believe our investigation provides the necessary background for a more comprehensive understanding of the engineering process, even though it has not examined more difficult engineering tasks like creating desired cellular structures using a number of co-occurring mutations or altering the cellular environment.

An essential goal of modern biotechnology is the examination of the link between genotype and phenotype with the goal of defining design principles for the rational engineering of cells and cellular structures. While it is crucial to obtain exact control over the design task, it is also crucial to perfect the experimental realisation. New professional figures will emerge as more computational methods are included into the bioengineering playbook, which will be advantageous to society as a whole. From an ethical standpoint, the need to produce desired results and consider a wider range of potential outcomes lessens the uncertainty surrounding the effects of the synthetic engineering process. Any pipeline that intends to create living things should have a more controlled design, even though this uncertainty cannot be totally eliminated from the equation.

3.9 CHROMOSOMAL ABNORMALITIES AND SYNDROMES

Almost every cell in the human body includes 23 chromosomal pairs, a total of 46 chromosomes. Half of our chromosomes are inherited from our mother, while the other half are inherited from our father. The initial 22 pairs are known as autosomes. The 23rd pair comprises the X and Y chromosomes, which are sex chromosomes. Females typically have two X chromosomes in each cell, while males have one X and one Y chromosome.

Chromosomal abnormalities come in many forms, but they can be categorised as numerical or structural. Whole chromosomes that are either extra or absent from their typical pair are considered numerical anomalies. Structural issues arise when a chromosome has a section that is missing, extra, moved to another chromosome, or inverted. Accidental chromosomal abnormalities can arise during the development of the egg, sperm, or early stages of the foetus. There may be a connection between the mother's age, the environment, and the frequency of genetic errors. Through prenatal screening and testing, it is feasible to examine the foetus' chromosomes and find some, but not all, types of chromosomal abnormalities.

3.9.1 Single gene disorders in humans

Chromosomes can be harmed or altered in number under particular conditions, such as those brought on by radiation from the environment, dietary factors, or inherited genetic abnormalities. Numerical chromosomal abnormalities refer to changes in the number, while structural chromosomal abnormalities (or aberrations) refer to changes in structure. When one of a pair of chromosomes is missing, the situation is referred to as monosomy (2n-1), for example, monosomy of chromosome 1. Trisomy (2n+1), such as that of chromosome X, is the term used to describe the presence of three copies of a chromosome. It's vital to note that aneuploidy, which encompasses both monosomy and trisomy, is a general term. However, the phenomenon is known as polyploidy when the full set of chromosomes is multiplied. Several polyploid plant varieties that are frequently used in our cuisine have been created through artificially breeding plants. For instance, tetraploid cabbages and mustards have four sets of chromosomes each, but bread wheat contains six sets (hexaploidy). Likewise, strawberries and sugar cane are octoploids, while bananas and apples both have three sets of chromosomes (8 sets of chromosomes). Significant phenotypic circumstances might change as a result of structural or numerical changes, manifesting as diseases or syndromes.

3.9.2 Structural chromosomal abnormalities

When a chromosome's structure or some components of it change, structural chromosomal abnormalities result. Typically, there are 46 chromosomes in total in each cell. If a chromosome has a portion that is missing, extra, or that has switched places with another portion, this is referred to as a structural chromosome anomaly. In the end, this causes either an abundance or a shortage of genetic material. This is a factor in several birth defects.

Structural chromosomal abnormalities may be of the following types:

1. **Deletion**— In deletion, a segment of a chromosome breaks away leading to the shortening of the chromosome (Fig. 3.9a). Retinoblastoma and Cri-du-chat syndrome, as examples. A part of chromosome 13 is deleted, which leads to retinoblastoma development.

2. **Duplication**— When a chromosome segment is duplicated, it lengthens the chromosome. This process is known as duplication (Fig. 3.9b). This may result in diseases like Charcot-Marie-Tooth disease, which is caused due to the duplication of certain genes on chromosome 17.

3. **Inversion**— An inversion occurs when a segment of a chromosome entirely separates, reverses, and then re-joins the chromosome. Here, the chromosome's overall length is the same, but the genes' orientation is 180 degrees backward (Fig. 3.9c). For instance, the inversion of a segment of chromosome 17 results in RCAD syndrome.

4. **Translocation**—A gene can be translocated from one linkage group to another through this process. Translocation occurs when a fragment of one chromosome separates and attaches itself to another chromosome. Reciprocal translocation is the term for the exchange of segments between two chromosomes. For instance, in Burkitt's lymphoma, the material is exchanged between chromosomes 8 and 14. Without mutual exchange, attachment is referred to as Robertsonian translocation. This can cause the cell's chromosome number to drop (Fig. 3.9d).

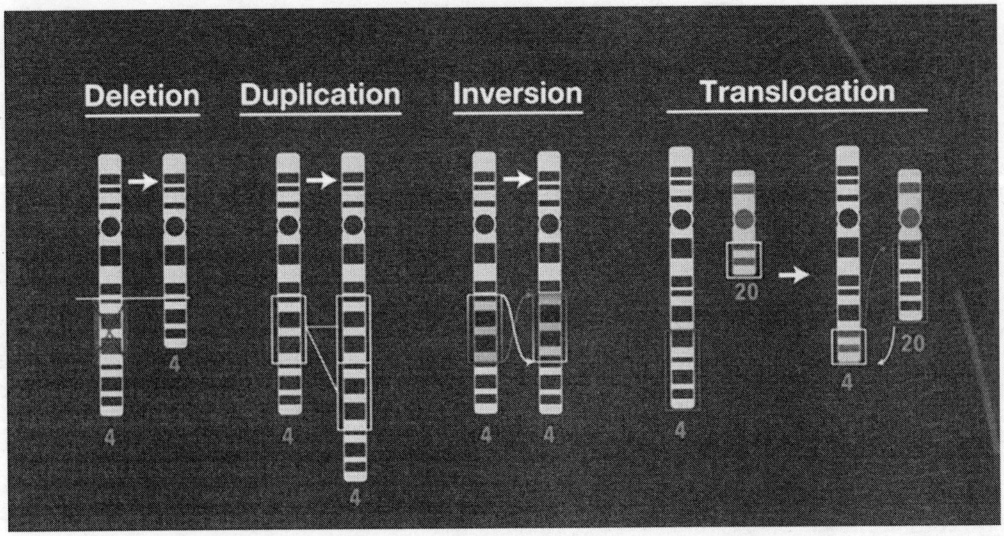

Fig. 3. 9 (a) Deletion (b) Duplication (c) Inversion (e) Translocation

(Source: https://www.genome.gov/about-genomics/fact-sheets/Chromosome-Abnormalities-Fact-Sheet)

3.9.3 Monogenic Disorders

An error in a single gene is the root cause of monogenic illness. According to current estimates, nearly 10,000 human diseases that impact millions of people globally are thought to be monogenic. The functions that the altered or faulty gene performs determine the type of disease, as well as its signs and symptoms. Mendel's Laws state that certain illnesses are inherited. The mutation may occur suddenly in some circumstances, in which case we won't know the preceding family history. One gene may have a single mutation producing a specific disease, such as sickle cell anaemia, or it may have several mutations producing the same disease, such as cystic fibrosis (more than 200 different types of mutation can occur in one gene). According to the inheritance pattern, single-gene or monogenic disorders can be divided into the following groups:

a) Autosomal recessive b) Autosomal dominant c) X-linked recessive d) X-linked dominant

a) Autosomal recessive disorder

Recessive means that two copies of the gene must be present for a trait or condition to exist in the case of a mutated gene. One gene out of the two copies is inherited from the father and the other from the mother. An individual will be the carrier and not get the disease if they have one normal recessive gene and one deficient recessive gene. According to statistical projection, it is thought that every human possesses at least five recessive genes that are defective and can lead to hereditary diseases. A recessive disorder's disease phenotype results from the homozygosity of a recessive allele, while the unaffected phenotype is caused by the matching dominant allele. This can be described using the example of the autosomal recessive disease sickle cell anaemia. A chromosome 11 haemoglobin gene mutation results in sickle cell disease. As a result, the haemoglobin is damaged (Hb). These faulty Hb molecules group together after donating their oxygen, forming rod-like formations.

Sickle cell anaemia is determined by an allele which we can designate as s and the normal condition by S. Affected individuals will have the genotype s/s, while unaffected individuals will either have S/S or S/s. People with sub-Saharan African, South American, Cuban, Central American, Saudi Arabian, Indian, and Mediterranean ancestry are particularly susceptible to sickle cell anaemia. It is widespread among residents of the Deccan plateau in central India, with a lesser concentration in Kerala and Tamil Nadu's northern regions.

Phenylketonuria, Tay Sach's disease, and cystic fibrosis are more examples of autosomal recessive illnesses. People with cystic fibrosis produce abnormally thick, sticky mucus that can harm many organs, particularly the lungs, leading to chronic infections. The absence of the hexosaminidase A enzyme, which causes Tay-Sachs disease, causes fatty material accumulation in nerve cells, with the brain being severely affected. It is a deadly condition that first appears in children. The Tay-Sachs gene is carried by one in every 27 individuals of European Ashkenazi Jewish descent. A phenylalanine hydroxylase gene mutation that results in an increase in blood phenylalanine is the cause of phenylketonuria.

b) Autosomal dominant disorder

The normal allele is recessive and the abnormal allele is dominant in this kind of inheritance. An uncommon autosomal dominant condition Achondroplasia is one condition that can cause a specific form of dwarfism in those who are affected. In this condition, those with mild disease have the genotype d/d, and those with severe disease have the genotype d/D, which is frequently fatal. Thus, heterozygotes make up the majority of the remaining instances of achondroplasia. Another rare autosomal dominant condition that affects the neurological system is Huntington's disease.

c) X-linked recessive disorder

Only males (XY) are typically afflicted by the disorder in X-linked recessive inheritance in the mother (XX), where the defective gene stays on one X chromosome. As a result, she becomes the carrier. The Y chromosome is passed down to sons while the X chromosome is passed down to daughters in the male progeny. A male who has the condition will thus not pass it on to his sons, but all of his daughters will be carriers. Haemophilia and Duchenne muscular dystrophy are a few X-linked recessive illnesses. Haemophilia is a bleeding disorder linked to mutations in the factor IX or VIII coagulation genes (type A) (type B). Coagulation factors VIII or IX are produced in an abnormal form or insufficiently as a result of mutations in the coagulation factor genes. Blood cannot properly clot due to the changed or absent coagulation factor, which results in increased or spontaneous bleeding tendencies. The dystrophin gene is mutated in Duchenne muscular dystrophy (DMD), which results in decreased or absent dystrophin or the presence of abnormal proteins. Muscles become more frail and weaker as a result of dystrophy or degeneration brought on by dystrophin abnormalities or deficiency.

d) X-linked dominant disorder

In this type of inheritance, the affected males pass on the mutated dominant gene to all their daughters but to none of their sons. In the case of affected females married to unaffected male, the condition is passed on to half of their sons and daughters. Examples of these conditions include Alport syndrome, which is linked to progressive hearing loss and renal disease, and hypophosphatemia, a kind of vitamin D-resistant rickets.

3.10 CONCEPT OF COMPLEMENTATION USING GENETICS

Complementation refers to a relationship between two different strains of an organism that both have homozygous recessive mutations that produce the same phenotype (for example, a change in wing structure in flies Fig. 3.10) but which do not reside on the same (homologous) gene. These strains have undergone true mutational breeding. When these strains are crossed, it is said that there has been "genetic complementation" if some of the offspring recover the wild-type phenotype. This results in heterozygous mutations in every linked gene in the progeny because each strain's haploid provides a wild-type allele to "complement" the mutant allele of the other strain's haploid. Due to the recessive nature of the mutations, the offspring will exhibit the wild-type phenotype.

A complementation test (sometimes called a "cis-trans" test) refers to this experiment, developed by American geneticist Edward B. Lewis. It responds to the query: "Does a wild-type copy of gene X rescue the function of the mutant allele thought to define gene X?" One can inquire as to whether the function that was lost as a result of the recessive allele can be replaced by another mutant genotype if there is an allele with an observable phenotype that can be provided by a wild-type genotype (i.e., the allele is recessive). If not, the same gene must have two defective alleles. The beauty of this test is that, even without understanding what the gene is doing at the molecular level, the trait may be used as a read-out of gene function.

Complementation develops because the same phenotype might result from the loss of function in genes responsible for various steps in the same metabolic pathway. When strains are crossed, the progeny receives either of the parent's wild-type copies of each gene. The recessive nature of the mutations causes a recovery of function in that pathway, which allows offspring to regain the wild-type phenotype. Consequently, the test is used to determine whether two independently derived recessive mutant phenotypes are brought on by mutations in one gene or two. If the same gene mutated in both parent strains, no normal copies of the gene are passed down to the offspring, who exhibit the same mutant phenotype, indicating that complementation has not taken place.

Fig 3.10 states that Due to two distinct autosomal recessive mutations that affect various steps in a single biochemical process that produces pigment, two strains of flies have white eyes. Since the progeny of their cross can complete the entire metabolic pathway and have red eyes as a result, flies from Strain 1 have complimentary mutations to flies from Strain. In other words, there are three possibilities if the fusion of two haploid genomes with various recessive mutations results in a mutant phenotype: Similar genes experience mutations; The expression of one mutation is impacted by the other's; A single mutation could produce an inhibiting substance. If two haploid genomes with different recessive mutations combine to produce the wild-type phenotype, the mutations must be in different genes.

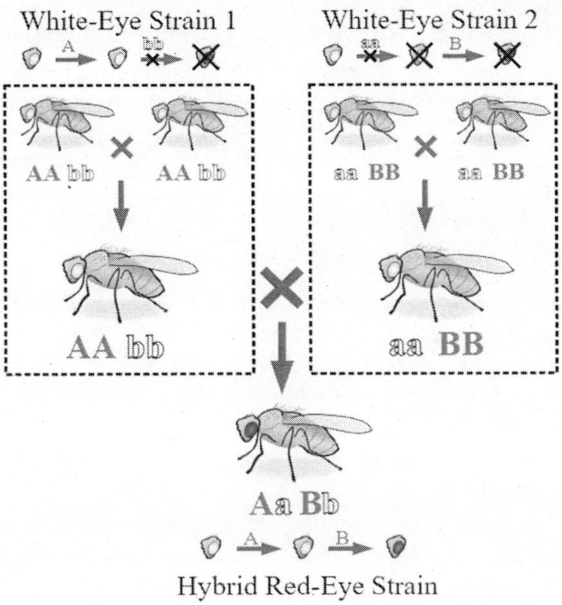

Fig. 3. 10 : Complementation Test: An illustration of a complementation test

(Source: https://en.wikipedia.org/wiki/Complementation_%28genetics%29) (Creative commons licenses)

In conclusion, genetics has always been concerned with how the hereditary information in DNA regulates the appearance and function of an organism. Historically, this consisted of using genetic variants (mutants) to disrupt the biological function of cells or animals and then deducing how cells and organisms functioned based on the effect of these mutations. At the molecular end of the subject, the availability of sequence information and genomic analysis, as well as sophisticated techniques for gene replacement and analysis of gene expression patterns (microarray technology), provides us with significantly more potent tools for examining how genes function to make us who we are. At the opposite end of the spectrum, genetic knowledge is essential to comprehending how organisms, populations, and species evolve. Through the application of the new molecular systematics to the challenges of development, evolution, and speciation, one of the most fascinating developments in the field over the past few years has been the approach of these two extremes.

Geneticists think that the tools and techniques of genetics are applicable across the entire spectrum of biological activity, and are as appropriate to molecular biology and population studies as they are to population genetics. Some of the fundamental tools of contemporary biology (analysis of genomic sequences and bioinformatics) are utilised most intelligently in the understanding of the genetic principles underlying the design and use of the software. On the opposite end of the spectrum, genetic knowledge is essential for comprehending the evolution of populations and species. The breakthroughs in sequencing genomes and microarray technology, which are now utilised by the vast majority of biologists, do not negate the fact that genetics offers a perspective and a variety of experimental methods applicable to numerous fields of biological investigation. To date, public health practice has focused on environmental or socioeconomic determinants of health and disease and has given genetic variability within the population limited consideration. The progress made in genomics is altering these perceptions. Long-term, this information will allow health promotion and disease prevention programmes to be focused particularly on susceptible individuals and families, or subsets of the community, depending on their genomic risk profile.

UNIT SUMMARY

- The three laws of inheritance proposed by Mendel include:
  - Law of Dominance - Hybrid offspring shows Dominant characters frequently and Recessive characters rarely.
  - Law of Segregation - 2 alleles of a gene always segregate during gamete formation
  - Law of Independent Assortment - The alleles segregate independently of each other during gamete formation
- Monohybrid cross - Observing the inheritance pattern with only one gene into consideration. The Monohybrid ratio is 3:1
- Dihybrid cross - Observing the inheritance of 2 different characters at a time. Ratio is 9:3:3:1
- Mutation: Any heritable change of the base-pair sequence of genetic material
- Allele - Each gene exists in two alternate forms called alleles (e.g; Height gene's allele - Tall & Short)
- Loci – The location of the gene on a chromosome is called a locus (Plural: loci). Always alleles of genes occupy the same loci
- Gene mapping - The linkage of the genes in a chromosome can be represented in the form of a genetic map
  - In the gene map, the distance is measured in terms of recombination frequency
  - Molecular markers enable us to identify the gen location and its linkage pattern
- Gene interaction - when two or more nonallelic genes influence the outcome of a single trait, this is known as Gene interaction
- Epistasis - is the interaction between different genes (i.e. non-alleles) whereas dominance is the interaction between different alleles of the same gene (i.e. intra-allelic)
- Mitosis is the type of cell division that results in the formation of two daughter cells each with the same number and kind of chromosomes as the parent cell. (e.g; skin cells)
- Meiosis is a type of cell division that results in the formation of four daughter cells each with half the number of chromosomes as the parent cell. (e.g; gametes)
- Any change in an organism's DNA (Chromosome) that is unique and heritable is termed 'Mutation'. Mutations are the sole reason for evolution and will remain the most important part of life
- Chromosomal abnormalities & Disorders: alteration can be structural or numerical
  - Deletion of a segment of a chromosome - Retinoblastoma
  - Duplication refers to when a segment of the chromosome gets repeated - Charcot Marie tooth disorder
  - In an inversion, a segment of the chromosome breaks away, completely reverses itself and reattaches with the chromosome

- In translocation, a segment of a chromosome breaks away and attaches itself with another chromosome
- Monogenic Disorders - caused by an error in a single gene
- Autosomal recessive disorder - Both allele copies of gene defected mutated causes this disorder - Sickle cell anaemia
- Autosomal dominant disorder – an inheritance of the normal allele is recessive and the abnormal allele is dominant - Huntington's disease
- X-linked recessive disorder - From mother, the affected gene remains on one X chromosome, as a result, she becomes the carrier and usually only males (XY) are affected - Haemophilia
- X-linked dominant disorder - In this type of inheritance the affected males pass on the mutated dominant gene to all their daughters but to none of their sons - Alport syndrome

- Complementation - a relationship between two different strains of an organism that both have homozygous recessive mutations that produce the same phenotype
- A complementation test allows us to determine whether two independently isolated mutants with the same phenotype have mutations in the same or different gene.

EXERCISES

Multiple Choice Questions

1) In a cross between a male and female, both heterozygous for the Sickle cell anaemia gene, what percentage of the progeny will be diseased?

A. 25%
B. 100%
C. 0%
D. 75%

2) Select the correct match

A. Phenylketonuria - Autosomal dominant trait
B. Sickle cell anaemia - Autosomal recessive trait
C. Hypophosphatemia - X linked
D. Haemophilia - Y linked

3) How many true-breeding pea plant varieties did Mendel select as pair that were similar except one character with contrasting traits?

A. 2
B. 14
C. 6
D. 4

4) The mechanism that causes a gene to move from one linkage group to another is called ______

A. Translocation
B. Crossing over

C. Duplication
D. None of the above

5) The incorrect statement with regard to haemophilia is ____

A. It is a recessive disease
B. A single protein involved in the clotting of blood is affected
C. It is a dominant disease
D. None of the above

6) Genotype of dominant plant can be determined by_____

A. Pedigree analysis
B. Back cross
C. Test cross
D. Dihybrid cross

7) The Phenomenon of two or more than two genes affecting the expression of each other is called _______

A. Crossing over
B. Pairing
C. Linkage
D. Gene interaction

8) _________ is a form of cell division which results in the creation of gametes or sex cells.

A. Mitosis
B. Meiosis
C. Miosis
D. None of the above

9) Continuous variations are due to ______

A. Mutation
B. Crossing over
C. Polyploidy
D. Chromosomal aberrations

10) "Cri-du-chat" syndrome is caused by change in a chromosome structure involving _____

A. Deletion
B. Duplication
C. Inversion
D. Translocation

11) The "cis-trans" test developed by American geneticist _____

A. Thomas Morgan
B. William Gosset
C. Karl Pearson
D. Edward Lewis

12) Which of the following is a result of reciprocal translocation?

A. Trichothiodystrophy
B. Burkitt's lymphoma
C. Cockyne's syndrome
D.
E. Thalassemia

13) RCAD syndrome caused due mutation in Gene HNF1B by inversion located on chromosome _____

A. 11
B. 15
C. 7
D. 17

14) The tendency of an offspring to resemble its parent is known as _____

A. Heredity
B. Variation
C. Resemblance
D. Inheritance

15) The geometrical test that helps predict the outcome of monohybrid or dihybrid crosses _______

A. Chi-square
B. Student T-test
C. Punnett square
D. ANOVA

16) When a recessive allele masks the expression of both dominant and recessive alleles, alleles referred to as what type of epistasis?

A. Dominant
B. Recessive
C. Duplicate-dominant
D. Duplicate recessive

17) Which of the following is NOT a type of epistasis?

A. Duplicate gene interaction
B. Complementary gene interaction
C. Polymeric gene interaction
D. Sex-linked

18) With four alleles, how many different combinations of alleles can there be?

A. 4
B. 8
C. 12
D. 16

Answers: 1) A; 2) B; 3) B; 4) A; 5) C; 6) C; 7) D; 8) B; 9) A; 10) A; 11) D; 12) D; 13) D; 14) A; 15) C; 16) A; 17) B; 18) C

Short Answer Type Questions

1. What do we inherit from our parents? Genotype or Phenotype?
2. What do you think, do we find Mendelian inheritance in Human beings too?
3. What do you think, parents, donate genes to offspring in equal numbers?
4. Chromosomal mutation and gene more lethal gene mutation chromosome number increases mutation rate increase, is the statement true?

NUMERICAL PROBLEM

A. If gene A and B genes are linked by a 10 cm distance. What might be the probability of obtaining the homozygous gametes for genes A and B?

KNOW MORE

- Paweletz, N. (2001), Walther Flemming: Pioneer of mitosis research. *Nature Reviews Molecular Cell Biology* **2**, 72–75 doi:10.1038/35048077
- Bateson, W. (1909). Mendel's Principles of Heredity: Cambridge University Press. *März 1909; 2nd Impr*, *3*, 1913.
- https://www.genome.gov/genetics-glossary/Epistasis
- https://www.nature.com/scitable/topicpage/epistasis-gene-interaction-and-phenotype-effects-460/
- https://www.ncbi.nlm.nih.gov/books/NBK19726/

REFERENCES AND SUGGESTED READINGS

- https://thebiologynotes.com/mendels-law-of-segregation/
- https://humanbiology.pressbooks.tru.ca/chapter/5-9-mendels-experiments-and-laws-ofinheritance/
- https://www.teacherspayteachers.com/Product/Genetics-Gregor-Mendel-Mendels-Laws-with-Worksheet-1811886
- https://www.ncbi.nlm.nih.gov/books/NBK115545/
- https://www.khanacademy.org/test-prep/mcat/behavior/behavior-and-genetics/a/genes-environment-and-behavior
- https://biologydictionary.net/epistasis/
- https://www.nature.com/scitable/topicpage/mitosis-meiosis-and-inheritance-476/#
- https://www.biorxiv.org/content/10.1101/2022.03.17.484826v1
- https://bio.libretexts.org/Bookshelves/Microbiology/Book%3A_Microbiology_(Boundless)/7%3A_Microbial_Genetics/7.11%3A_Genetic_Transfer_in_Prokaryotes/7.11E%3A_Complementation

Dynamic QR Code for Further Reading

4 Biomolecules

UNIT SPECIFICS

Through this unit, we have discussed the following aspects:

- *The Molecules of Life- Biomolecules*
- *Understanding that macromolecules like proteins and nucleic acids are polymers of monomers with distinct chemical properties*
- *Carbohydrates as a fuel of living organisms*
- *Nucleic acids as the carrier of genetic information*
- *Lipids and Fats as store-house of energy*

The ***practical applications of the topics are discussed*** *for generating further curiosity and creativity as well as improving problem solving capacity.*

Besides giving a large number ***of multiple choice questions as well as questions of short and long answer types*** *marked in two categories following lower and higher order of* ***Bloom's taxonomy****, assignments through a number of numerical problems, a list of references and suggested readings are given in the unit so that one can go through them for practice. It is important to note that for getting more information on various topics of interest some* ***QR codes*** *have been provided in different sections which can be scanned for relevant supportive knowledge.*

After the related practical, based on the content, there is a ***"Know More"*** *section. This section has been carefully designed so that the supplementary information provided in this part becomes beneficial for the users of the book. This section mainly highlights the initial activity, examples of some interesting facts, analogy, history of the development of the subject focusing the salient observations and finding, timelines starting from the development of the concerned topics up to the recent time, applications of the subject matter for our day-to-day real life or/and industrial applications on variety of aspects, case study related to environmental, sustainability, social and ethical issues whichever applicable, and finally inquisitiveness and curiosity topics of the unit.*

RATIONALE

To convey that all forms of life has the same building blocks and yet the manifestations are as diverse as one can imagine. Molecules of life. In this context discuss monomeric units and polymeric structures. Discuss about sugars, starch and cellulose. Amino acids and proteins. Nucleotides and DNA/RNA. Two carbon units and lipids.

It is crucial for engineering students to comprehend the fundamentals of engineering and the introduction of biological concepts in order to interact well on finding solutions to issues relating to biosystems. Learning this will help to apply engineering principles for analysing biological systems and develop energy-saving technologies that work in harmony with biosystems.

PREREQUISITES

Biology: Biomolecules (Class XI)

Chemistry: Biomolecules (Class XII)

UNIT OUTCOMES

List of outcomes of this unit is as follows:

U4-O1: Describe the role of biomolecules in the biosystem

U4-O2: List the four major classes of macromolecules and distinguish between monomers and polymers

U4-O3: Discuss structure, location and function of Glucose, Cellulose and Starch
Explain classification of carbohydrates on basis of hydrolysis

U4-O4: Differentiate DNA and RNA and discuss stability of DNA and RNA

U4-O5: Describe structure and functions of lipids and fats

Unit-4 Outcomes	**EXPECTED MAPPING WITH COURSE OUTCOMES** (1- Weak Correlation; 2- Medium correlation; 3- Strong Correlation)				
	CO-1	**CO-2**	**CO-3**	**CO-4**	**CO-5**
U4-O1	2	3	1	1	1
U4-O2	1	3	1	1	1
U4-O3	1	3	1	2	2
U4-O4	1	3	3	1	1
U4-O5	1	3	1	2	2

4.1 INTRODUCTION

A chemical reaction is initiated only when the molecules attain specific conditions (energy), appropriate complexity, and number. Although molecules and atoms don't have life, given the right energy flow and conditions, they can start a series of events that give the impression that they are alive. The molecules involved in organic processes and chemical reactions in living cells are identical. They can move, develop, execute the amazing chemistry of metabolism, react to environmental cues, and, most importantly, replicate themselves with astonishing reliability. Living organisms' intricate structures and behaviors hide the fundamental fact that they are made of molecules. Any molecules found in and produced by living organisms are called biomolecules. Biomolecules are crucial to the functioning of living organisms. Living organisms' molecular components do not randomly combine C, H, O, and N atoms in any of their infinite combinations. Instead, only a tiny proportion of the countless options are identified, and these categories have specific traits vital to the formation and sustenance of the living state. The sequence of escalating structural complexity by which macromolecular complexes are constructed from simple molecules is the main feature of biomolecular structure. The inorganic N-compounds ammonium salts (NH_4^+), nitrate salts (NO_3^-), and di-nitrogen molecules (N_2), together with water and carbon dioxide, are the primary precursors for the creation of biomolecules. These inorganic precursors are consumed and integrated by metabolic processes at complex levels of biomolecular structure. Biomolecules are classified into macromolecules (such as proteins, lipids, nucleic acids, and polysaccharides) and micromolecules (e.g., amino acids, vitamins, and monosaccharides). These compounds either carry out or initiate essential metabolic processes in living organisms. The biomolecules participate in a variety of processes, including the storage of energy (carbohydrates), the catalysis of biochemical reactions (enzymes), the storage and conveyance of genomic information (RNA/DNA), and the modification of biological and neurological functions (neurotransmitters/hormones). The study of biomolecules allows one to understand the physiological process that controls the proper growth and development of a human body.

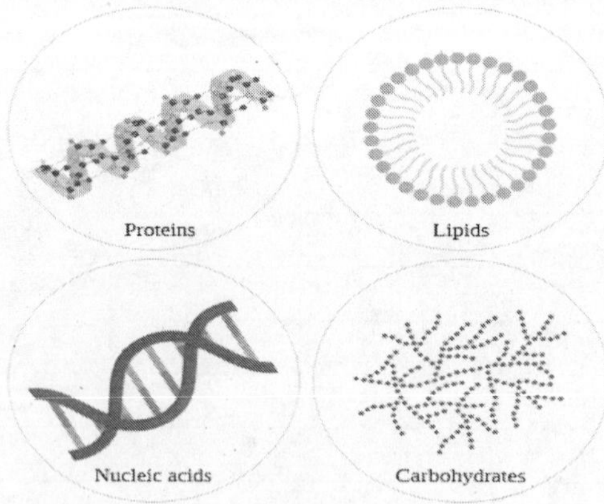

Fig. 4. 1 Biomolecules

(Images created using BioRender® software)

4.2 STRUCTURAL ORGANIZATION OF COMPLEX BIOMOLECULES

In the first phase, precursors are transformed into metabolites, which are basic organic chemicals that act as intermediaries in the biogenesis of several building block sets, such as glycerol, sugar complex carbohydrates, amino acids, nucleotides, and fatty acids. These building ingredients are joined covalently to form macromolecules, including proteins, polysaccharides, and polynucleotides (DNA and RNA). The next stage of structural organization, supramolecular complexes, is produced by interactions between macromolecules. Various members of one or more macromolecule classes combine to create particular assemblies that perform crucial subcellular tasks. Chromosomes, multifunctional enzyme complexes, ribosomes, and cytoskeletal components are examples of supramolecular assemblies. Before going into these complex structures, we shall have an in-depth view of these monomer molecules and then the polymeric structures of these monomers.

4.2.1 Monomers & Polymers

Polymers are compounds consisting of several smaller monomeric subunits that are bonded covalently. Condensation is a chemical process that binds interacting biological molecules. The removal of water creates a molecule of water for each chemical link, wherein -H and -OH atoms are extracted from opposite sides, forming a bond between the two molecules. In contrast to condensation, hydrolysis releases smaller molecules by rupturing the polymer's bond(s) (often individual monomers). To break the connection, a water molecule is needed. Doing so adds -H and -OH to either side.

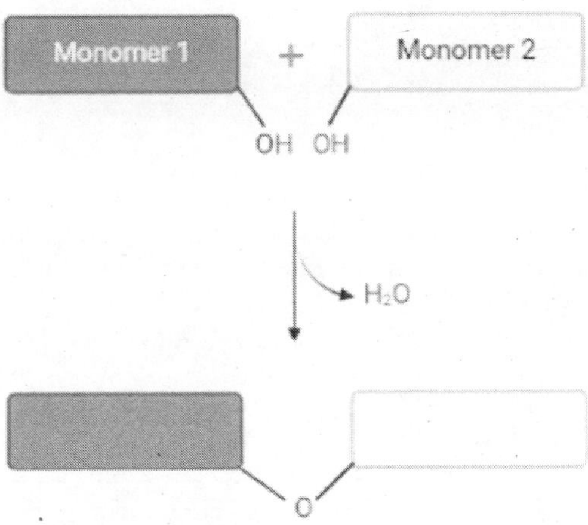

Fig. 4. 2 Condensation reaction

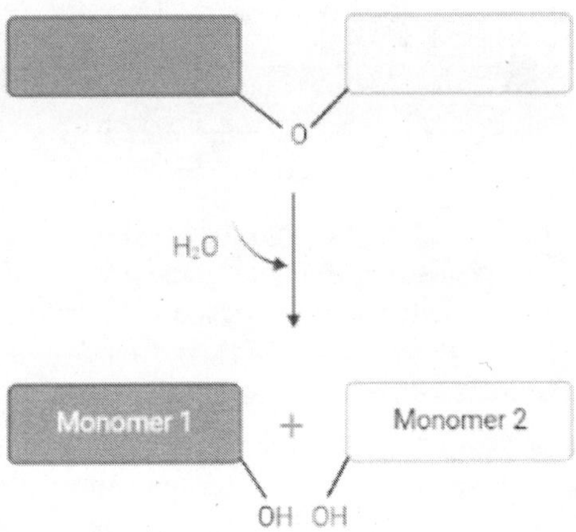

Fig. 4. 3 Hydrolysis reaction

(Images created using BioRender® software)

4.3 PROTEINS

4.3.1 Amino acids

Amino acids are the building blocks that form polypeptides and, ultimately, proteins. Consequently, they are fundamental components of our bodies and vital for physiological functions such as protein synthesis, tissue repair, and nutrient absorption. As demonstrated in Fig. (4.4). The fundamental structure of amino acids is the same for all of them and consists of a core carbon atom coupled to:

1) An amine group (NH_2)	2) A carboxylic acid group (COOH)
3) A hydrogen atom (H)	4) A variable side chain (R)

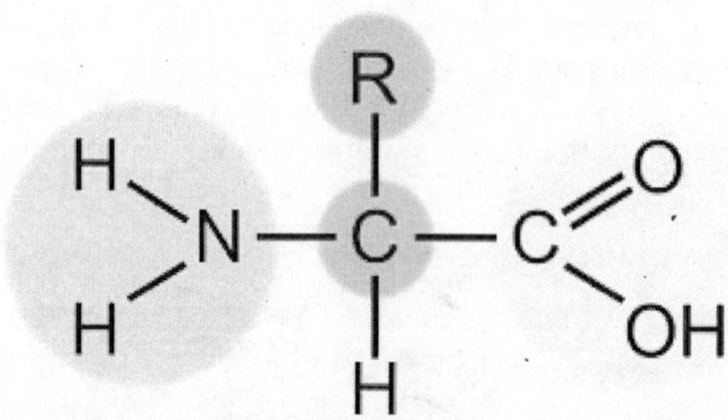

Fig. 4. 4 Empirical structure of amino acids

Source: wikimedia.org (Creative Common License)

Proteins include 22 different amino acids, although only 20 of them are listed in the human genetic code. Pyrrolysine and selenocysteine are two uncommon amino acids that are found in nature. Because they are less common in nature than the other amino acids, they are known as rare amino acids. While Pyrrolysine (Pyl) is found in methanogenic archaea, Selenocysteine (Sec) is prevalent in bacteria and eukaryotes.

Classification of Amino acids on basis of Polarity - The structure of the variable side chain varies for each kind of amino acid. According to their exact positions within the polypeptide chain, these side chains will have varied chemical characteristics (e.g., charged, non-polar, etc.), causing the protein to fold and function differently.Amino acids are grouped into five categories based on their polarity: (Fig 4.5).

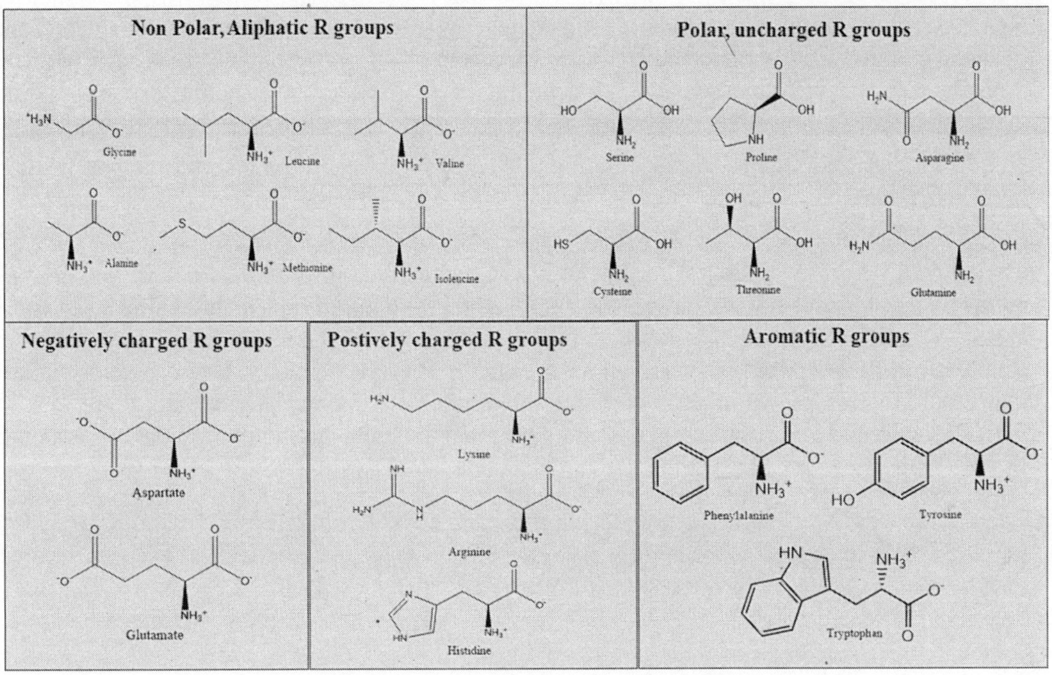

Fig. 4. 5 Amino acids classified on the basis of polarity

(Source: Created with ChemDraw software)

Non-Polar Amino Acids (Hydrophobic) - The aliphatic and aromatic amino acids are two different distinct types into which the nonpolar amino acids can be broadly classified. Glycine, alanine, valine, leucine, isoleucine, and proline are aliphatic amino acids that often have branching hydrocarbon chains. Because of the high carbon/hydrogen content and the presence of aromatic functional groups in their structure, the aromatic amino acids (phenylalanine, tyrosine, and tryptophan) are predominantly nonpolar and hydrophobic.

Polar Amino Acids (Hydrophilic) - The polar, hydrophilic amino acids are classified into three broad categories: polar uncharged-, acidic-, and basic-functional groups. The side chains of the polar uncharged class contain heteroatoms (O, S, or N) that can create long-lasting dipoles inside the R-group. These comprise amide-containing amino acids and hydroxyl- and sulfonyl-containing amino acids (serine, threonine, and cysteine) (glutamine and asparagine). The side chains of the acidic amino acids with carboxylic acid functional groups include glutamic acid and aspartic acid, which can be completely ionized in solution. Lysine, arginine, and histidine are three basic amino acids that have amine functional groups that can be protonated to carry a complete charge.

Classification of amino acids based on nutrition - Amino acids are classified into three categories:

Essential Amino acids: The term "essential amino acids" refers to amino acids the body cannot produce on its own. Of the 20 amino acids, 9 are regarded as being necessary. Essential amino acids must be obtained from diets such as soy, quinoa, egg, chicken, meat, or vegetable protein. There are nine essential amino acids: histidine, leucine, phenylalanine, valine, tryptophan, threonine, isoleucine, methionine, and valine.

Non-Essential Amino acids: Of the total of twenty amino acids, eleven are considered non-essential amino acids since the body may produce them on its own. Alanine, arginine, asparagine, aspartic acid, cysteine, glutamic acid, glutamine, glycine, proline, serine, and tyrosine are among the non-essential amino acids.

Conditional Amino acids: Because the body cannot synthesize enough arginine and histidine during particular physiological times of growth, such as pregnancy, teenage growth, or recovery from trauma, these amino acids may be regarded as conditionally necessary.

Stereochemistry of Amino Acids - All other amino acids included in protein structures are chiral in nature except for glycine, the most basic amino acid. All naturally occurring chiral amino acids are S, except for Cys, which is R, according to modern stereochemistry assignments using the Cahn-Ingold-Prelog priority criteria, which are widely used in chemistry. L-amino acids make up almost all proteins known to exist in plants and animals. However, D-amino acids are found in the cell walls of some bacteria, and some antibiotics (such as actinomycin D and gramicidin) include various levels of D-leucine, D-phenylalanine, and D-valine.

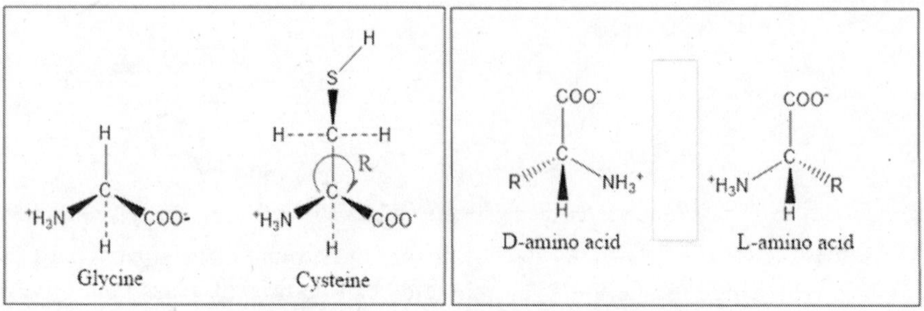

Fig. 4. 6 Stereochemistry of Amino acids

(Source: Created with ChemDraw software)

4.3.2 Peptide Bond Formation and Primary Protein Structure

The carboxylic acid of the upstream amino acid and the amine functional group of the downstream amino acid combine to produce an amide linkage, which is how a protein's main sequence is joined together by a dehydration synthesis (loss of water). The opposite reaction, called hydrolysis, requires the inclusion of a water molecule to break the amide bond and separate the two amino acids. It should be noted that the ribosome acts as the enzyme that conducts the dehydration synthesis processes necessary to construct protein molecules. In contrast, a class of enzymes called proteases is essential for protein breakdown. The amide linkage between amino acids inside protein structures is a peptide bond. More amino acids will be added to the carboxylic acid terminal of the developing protein. As a result, the amine and carboxylic acid tails are always produced in a particular order throughout the synthesis of proteins. The carboxylic acid tail of the previous amino acid is always where new amino acids are introduced, never the amine. The ribosome controls the directionality of N- to C- synthesis, which is how proteins are synthesized.

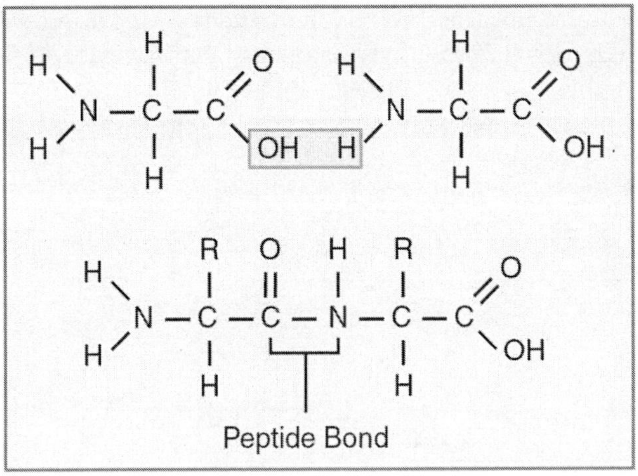

Fig. 4. 7 Formation of the Peptide Bond

Source: wikimedia.org (Creative Common License)

Proteins are significant molecules that consist of several amino acid residues connected in a very specific manner. Proteins come in various sizes, with the longest one known to exist having 33,423 amino acids named Titin. Peptides are macromolecules containing less than 50 amino acids, one of the smallest known human proteins, thymosin, only has 44 amino acids. Based on the protein's environment, a protein's folding pattern is determined by the kind and order of the amino acids in its main sequence (i.e., if it is inside the cell, it is likely surrounded by water in a very polar environment, whereas if the protein is embedded in the plasma membrane, it will be surrounded by very nonpolar hydrocarbon tails). Many alternative protein combinations may be employed to produce unique protein structures since a vast selection of amino acids can be inserted at each place inside the protein. Consider a tripeptide created from this pool of amino acids as an illustration. Twenty distinct alternatives may be used at each location. This means that there are a total of 20^3 tripeptides that might result, which is equal to 8,000 potential combinations. Think about the potential uses for a 40-amino-acid short peptide. 20^{40} possibilities, or a staggering 1.09 X 10^{52} possible sequence alternatives, would be available. Since the nature of the amino acid side chains influences how the protein interacts with other residues within the protein and its environment, each of these alternatives would result in a different overall protein shape.

The characteristics of the amino acids that make up a protein's 3-dimensional structure aid in protein folding. Protein form equals protein function. Hence the protein must have this 3-D structure to function. Hydrophobic amino acids are frequently located on the interior of protein structures for proteins found within the watery surroundings of cells. In contrast, hydrophilic amino acids, which love water, are found on the surface, where they may form hydrogen bonds and interact with the water molecules. Because it is the sole R-group that associates with the amine functional group in the main chain to generate a cyclic structure, proline is special. Proline adopts the cis conformation inside the backbone rather than the trans conformation due to this cyclization. Since the protein's structure has changed, prolines are frequently seen at locations where bends or directional changes take place. As the first amino acid for practically all of the many thousands of proteins known to exist in nature, methionine is special. Cysteine can be oxidized with other cysteine residues to create covalent disulfide connections within the protein structure since they have thiol functional groups (Figure 2.14). Disulfide bridges increase the 3-D structure's stability and are frequently necessary for proper protein folding and function.

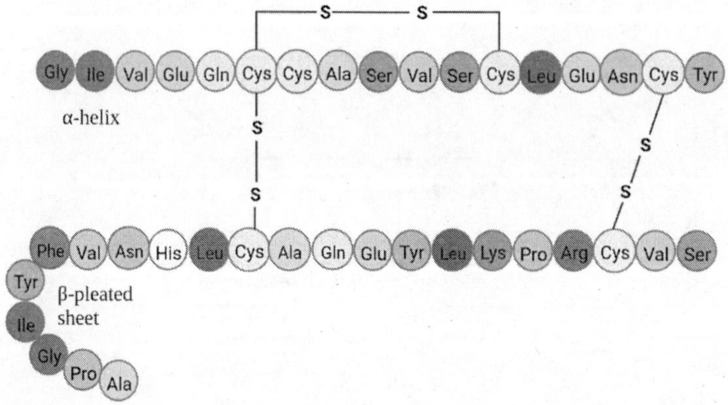

Fig. 4. 8 Disulfide Bonds

(Images created using BioRender® software)

4.3.3 Protein Shape and Function

The foundation of a protein's interaction with other molecules, 0tertiary, and occasionally quaternary protein structures in Module 7.

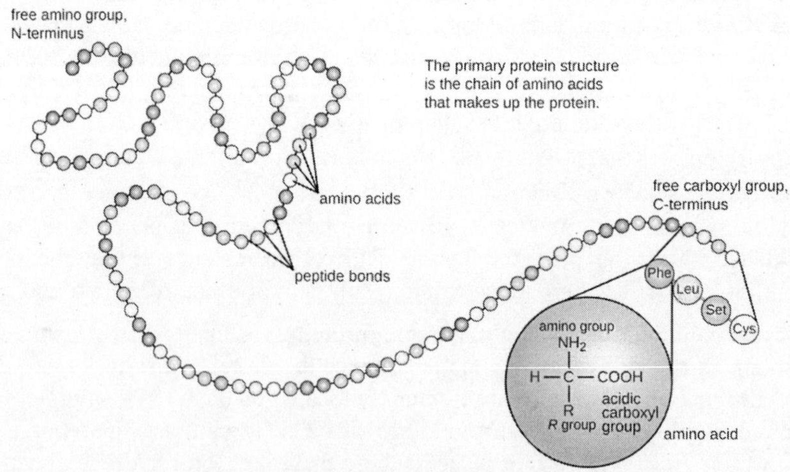

Fig. 4. 9 Primary protein structure

Source: wikimedia.org (Creative Common License)

4.4 CARBOHYDRATES

Carbohydrates are represented by the chemical formula $(CH_2O)_n$,where n is the number of carbon atoms in the molecule. Carbon, hydrogen, and oxygen are arranged in the molecules of carbohydrates in a 1:2:1 ratio. Thus, the definition of the word "carbohydrate" is explained by the fact that it is composed of carbon ("carbo") and water ("hydrate"). Ketones or polyhydroxy aldehydes can be referred to as carbohydrates. Monosaccharides, disaccharides, and polysaccharides are the three subcategories of carbohydrates.

Classification of Carbohydrates

Monosaccharide	Disaccharide	Polysaccharide
• Glucose	• Maltose (Glucose +Glucose)	• Starch (α Glucose)
• Fructose	• Sucrose (Glucose+Fructose)	• Cellulose (β Glucose)
• Galactose	• Lactose(Glucose+Galactose)	• Glycogen(α Glucose)

4.4.1 Monosaccharides

Monosaccharides, commonly referred to as simple sugars, are the most basic kind of carbohydrates. They have the general formula $C_n(H_2O)_n$and cannot be further hydrolyzed. The monosaccharides are divided into distinct categories based on the functional group and the number of carbon atoms. They are divided into two categories based on functional groups: aldoses and ketoses.

Aldoses: When the functional group in monosaccharides is an aldehyde, they are known as aldoses, e.g. glyceraldehyde, and glucose.

Ketoses: When a functional group is a keto group, they are called ketoses e.g,. dihydroxyacetone, and fructose.

Monosaccharides

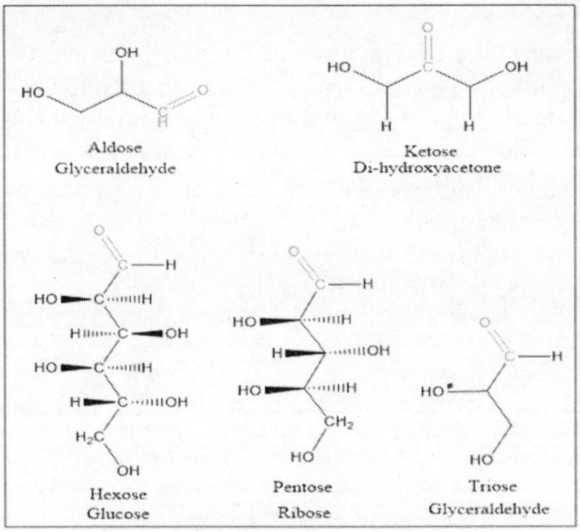

Fig. 4. 10 Monosaccharides

(Source: Created with ChemDraw software)

The monosaccharides are classified as trioses (3C), tetroses (4C), pentoses (5C), hexoses (6C), and heptoses (7C) based on the number of carbon atoms. When naming monosaccharides, these terms are used with functional groups. For example, fructose is a ketohexose, whereas glucose is an aldohexose.

Monosaccharides can take the form of a linear chain or a ring-shaped molecule. In aqueous solutions, they are often found in ring shapes. When glucose is in a ring shape, the hydroxyl group (OH) can be arranged in one of two ways around the anomeric carbon (carbon 1 becomes asymmetric during ring formation). When the hydroxyl group is found below carbon atom number 1, it is said to be in the alpha position (α), and when it is located above the plane, it is said to be in the beta position (β).

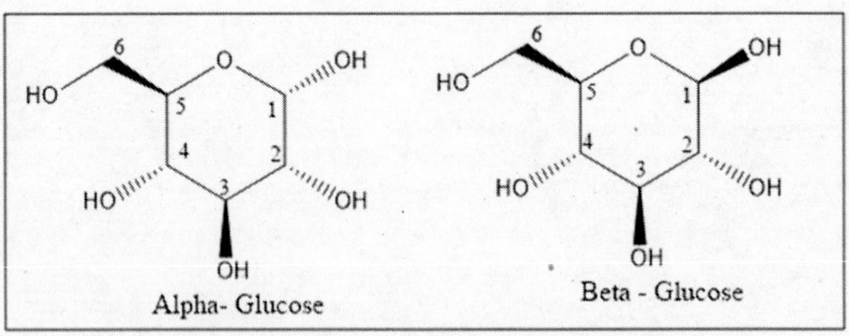

Fig. 4. 11 Stereochemistry of glucose

(Source: Created with ChemDraw software)

4.4.2 Oligosaccharides

There are 2 to 10 monosaccharides in oligosaccharides. Glycosidic linkages, or covalent bonds, hold the monosaccharides together. A condensation process between two hydroxyl (-OH) groups between any two monosaccharides results in the formation of a glycosidic linkage.

4.4.3 Types of linkages

α- and β-glycosidic bonds - The relative stereochemistry of the anomeric location and the stereocenter farthest from C1 in the saccharide can be used to differentiate between α- and β glycosidic linkages when an anomeric center is present.

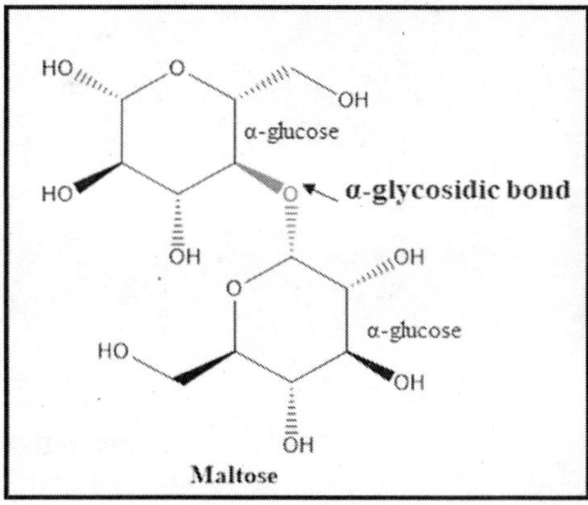

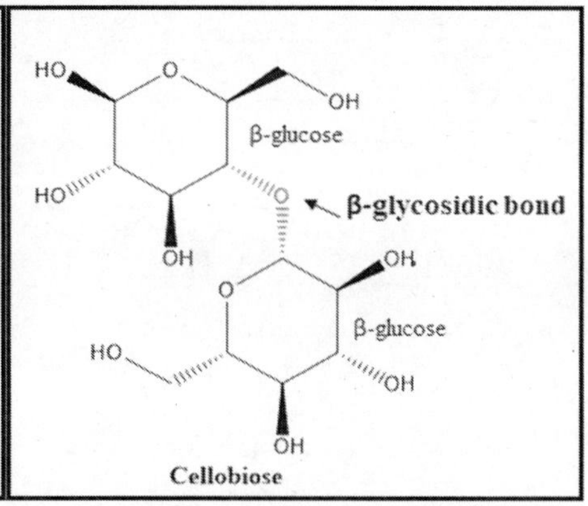

Fig. 4. 12 : α-glycosidic linkage in maltose

Fig. 4. 13 β-glycosidic linkage in Cellobiose

(Source: Created with ChemDraw software) e

1,4-glycosidic bond - A covalent link between the -OH group on carbon 1 of one sugar and the -OH group on carbon 4 of another sugar is known as a 1,4-glycosidic linkage.

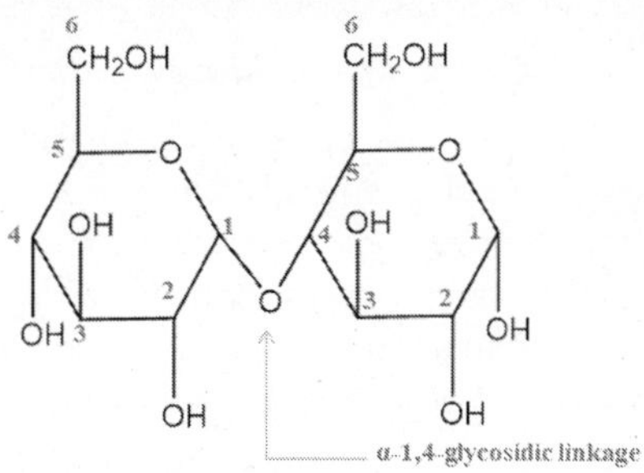

Fig. 4. 14 α-glycosidic linkage in Maltose (2 α-glucose monomers bind via α-glycosidic linkage to form Maltose)

(Source: Created with ChemDraw software)

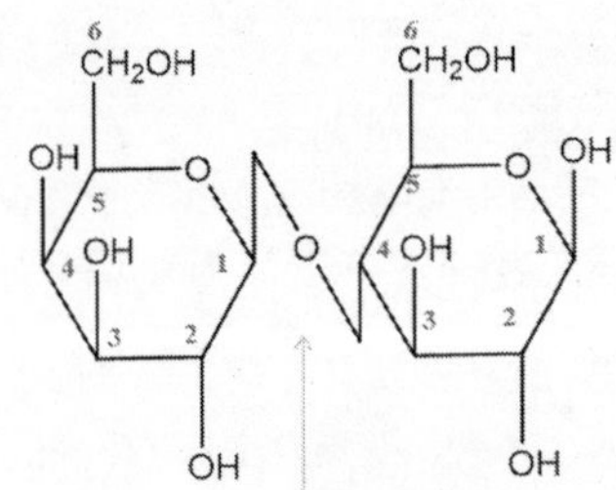

Fig. 4. 15 β-glycosidic linkage in Maltose (β-glucose & β-galactose binds via β-glycosidic linkage to form Maltose)

(Source: Created with ChemDraw software)

1,6-glycosidic bond - A covalent link between the -OH group on carbon 1 of one sugar and the -OH group on carbon 6 of another sugar is known as a 1,6-glycosidic bond. The polysaccharide branches due to this linkage.

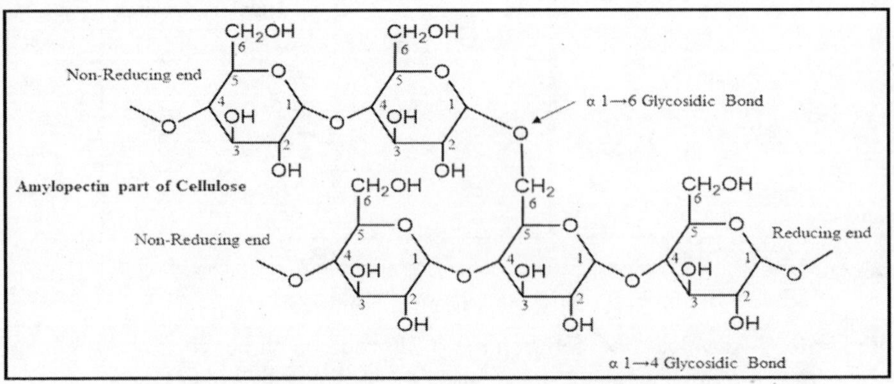

Fig. 4. 16 1,6-glycosidic bond in amylopectin

(Source: Created with ChemDraw software)

The oligo-saccharides are further split into disaccharides, trisaccharides, etc., depending on the number of monosaccharide units present. A **disaccharide** comprises two monosaccharide units, which can be the same or different, and is kept together by a glycosidic bond. Distinct features include crystalline nature, water-solubility, and sweetness to taste. Examples of disaccharides include lactose, maltose, and sucrose, typically found in nature and everyday diets. The monomers of lactose are glucose and galactose. They naturally occur in milk. Maltose, often known as malt sugar, is a disaccharide formed when two molecules of glucose undergo a condensation reaction in which a water molecule is lost. Yet another prevalent disaccharide is sucrose, or table sugar, made of fructose and glucose.

Two forms of disaccharides exist:

a. Reducing sugars: If the carbonyl group in the sugars is not involved in the glycosidic linkage, then they retain their reducing property; such sugars are called reducing sugars, i.e., having a free aldehyde or ketogroup end, e.g., maltose, lactose.

b. Non-reducing sugars: If the carbonyl group in the sugars is involved in the glycosidic linkage, then they are not available for reduction. Such sugars are callednon-reducing sugars, i.e. have NO free aldehyde or ketogroup end, e.g., sucrose, trehalose.

4.4.3 Polysaccharides

Polysaccharides are high-molecular-weight polymers comprising monosaccharide molecules (up to a million). They typically have no taste (non-sugars) and combine with water to form colloids. There are two different kinds of polysaccharides: homopolysaccharides and heteropolysaccharides.

A homopolysaccharide is a polysaccharide that includes the same type of monosaccharide. Glycogen, Cellulose, and Starch are examples of essential homopolysaccharides.

A heteropolysaccharide is a polysaccharidethat includes different types of monosaccharides. Hyaluronic acid and Heparin are examples of homopolysaccharides.

Table 4. 1 Classification of carbohydrates

	Disaccharides			Polysaccharides		
Carbohydrates	Maltose	Sucrose	Lactose	Starch	Cellulose	Glycogen
Linkages	α-1,4-glycosidic linkages	α-1,2-glycosidic linkages	β-1,4-glycosidic linkages	α-1,4-glycosidic linkages & α-1,6-glycosidic linkages	β-1,4-glycosidic linkages	α-1,4-glycosidic linkages & α-1,6-glycosidic linkages
Monomers	Glucose	Glucose & Fructose	Glucose & Galactose	α-Glucose	β-Glucose	α-Glucose

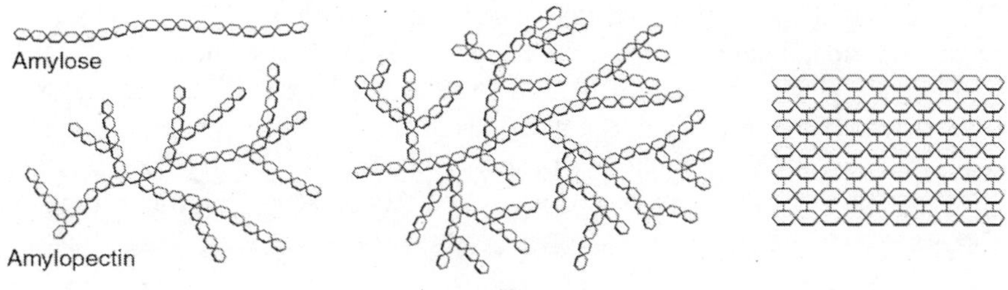

Fig. 4. 17 Difference in the arrangement of glucose monomers in different polysaccharides

Source: wikimedia.org (Creative Common License)

Cellulose - A straight, ribbon-shaped polymer of glucose molecules called cellulose with the C1, and C4 carbons of the glucopyranose rings form glycosidic connections to connect the individual glucose units. Cellulose is a tough, water-insoluble, and fibrous polysaccharide. About 30% of the plant cell wall is composed of cellulose. It is one of the main elements in plant cell walls that make plants stiff and strong. To sustain the cell walls of plants, cellulose fibers are embedded in a polysaccharide matrix. Plant stems and wood get support from cellulose fibers distributed in a lignin matrix. Cotton, which has a cellulose content of more than 90%, is the purest natural type of cellulose. Wood, on the other hand, is 40–50% cellulose. It differs from starch because of its beta-acetal linkage. Human digestibility varies significantly due to these specific variations in acetal connections. Humans lack the necessary enzymes to break the beta-acetal bonds, and therefore cellulose cannot be digested by them. The intestinal tract of many animals, including termites, cows, horses, sheep, and goats, contains symbiotic bacteria. They possess the necessary enzymes for the breakdown or hydrolysis of cellulose. No vertebrate can directly digest cellulose. Although humans cannot digest cellulose, it is still used in various items, such as films, cotton, linen, rayon for clothing, paper products, nitrocellulose for explosives, and wood for construction.

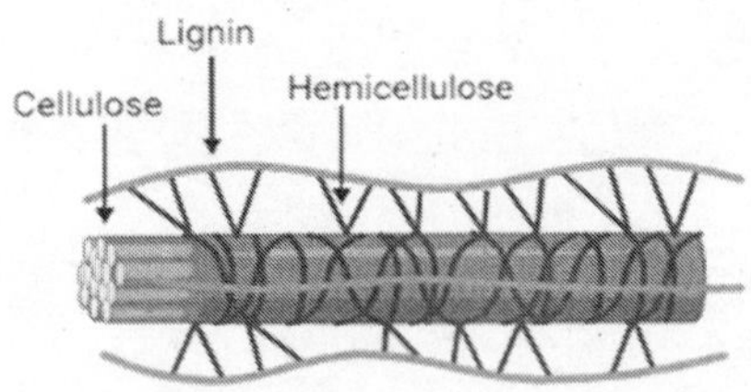

Fig. 4. 18 Plant Cellulose

(Images created using BioRender® software)

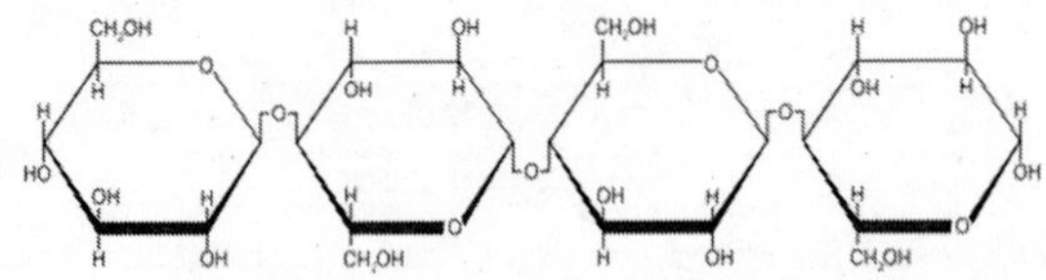

Fig. 4. 19 Structure of cellulose

Source: wikimedia.org (Creative Common License)

Starch - The chloroplasts of green leaves and the amyloplasts of tubers and grains contain starch, which is semi-crystalline in nature. Amylose and amylopectin are the two polysaccharide units that constitute starch chemically. Amylopectin is crystalline in form, whereas amylose is amorphous. Glucose units make up both amylose and amylopectin. Amylopectin is a branched chain polymer having α -d-(1-4) glycosidic links in the linear chain and α -d-(1-6) branched linkages, whereas amylose is a linear polymer connected by α -d-(1-4) glycosidic linkages, making up around 20–30% of total starch. The structural basis for the semi-crystalline nature of starch is provided by amylopectin. Amylose is much more compact and has fewer branches. The starch granules are thought to become denser because it is thought to fill any gaps in the semi-crystalline matrix created by amylopectin. Each plant species produces different types of starch with unique components (such as the ratio of amylose to amylopectin or the length of glucose chains), and the protein and fat content of the storage organs might differ greatly. Thus, depending on the source, starch varies.

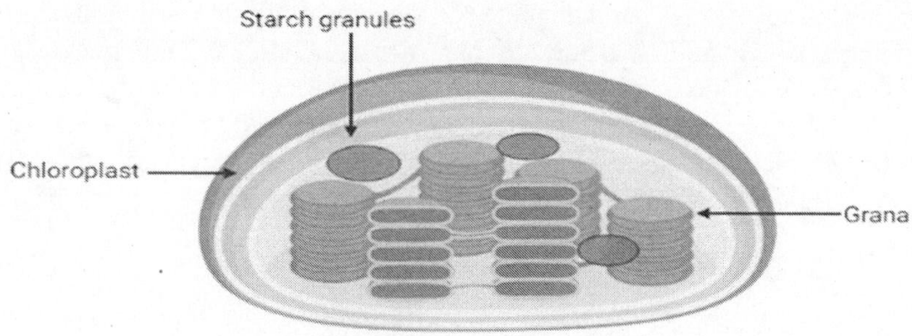

Fig. 4. 20 Starch in plants

(Images created using BioRender® software)

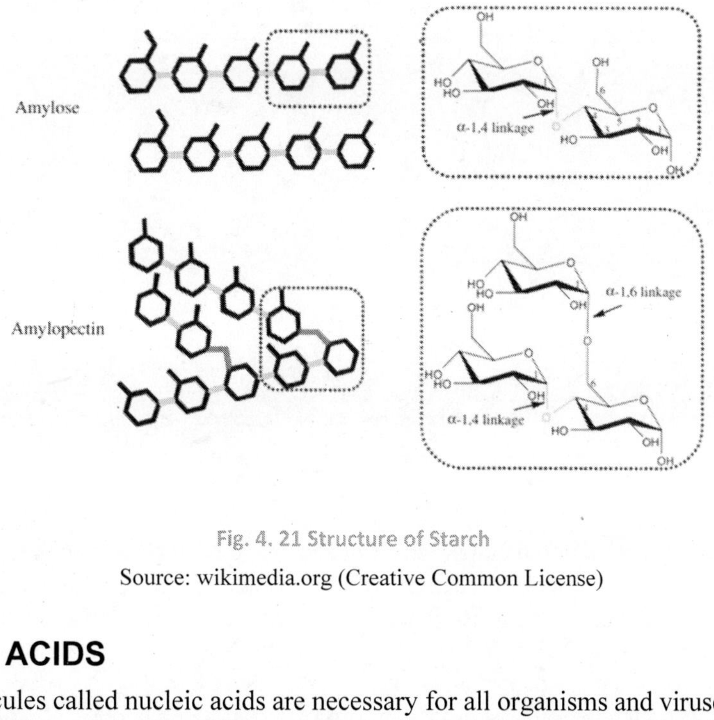

Fig. 4. 21 Structure of Starch

Source: wikimedia.org (Creative Common License)

4.5 NUCLEIC ACIDS

Large macromolecules called nucleic acids are necessary for all organisms and viruses to function. The templates for the proteins that cells produce are contained in nucleic acids, which are information molecules. The preservation and expression of genetic data is a crucial role of nucleic acids. Additionally, because reproducing cells convey the blueprints to their progeny, they serve as the genetic material in cells. Deoxyribonucleic acid (DNA) and ribonucleic acid are the two primary forms of nucleic acids (RNA). Nucleotides are the monomers that make up DNA and RNA. A nucleic acid, such as DNA or RNA, is created when the nucleotides combine. Each nucleotide has a nitrogenous base, a pentose (five-carbon) sugar, and a phosphate group, whereas nucleosides only have a nitrogenous base and a five-carbon carbohydrate group. Consequently, a nucleoside with one or more phosphate groups attached is what is known as a nucleotide. The nitrogenous base may be either a purine or a pyrimidine. Purines are two-ring structures that include inosine (I), adenine (A), and guanine (G) whereas pyrimidines are one-ring structures that include uracil (U), cytosine (C), and thymine (T) (Fig. 4.22). Both purine and pyrimidine nitrogenous bases are made from amino acids. Nucleotide's nitrogenous bases are each joined to sugar molecules that have one or more phosphate groups linked to them. Adenine (A), Thymine (T), Guanine (G), and Cytosine (C) are the bases that are employed in DNA. Thymine (T) is replaced by the nucleotide Uracil (U) in RNA.

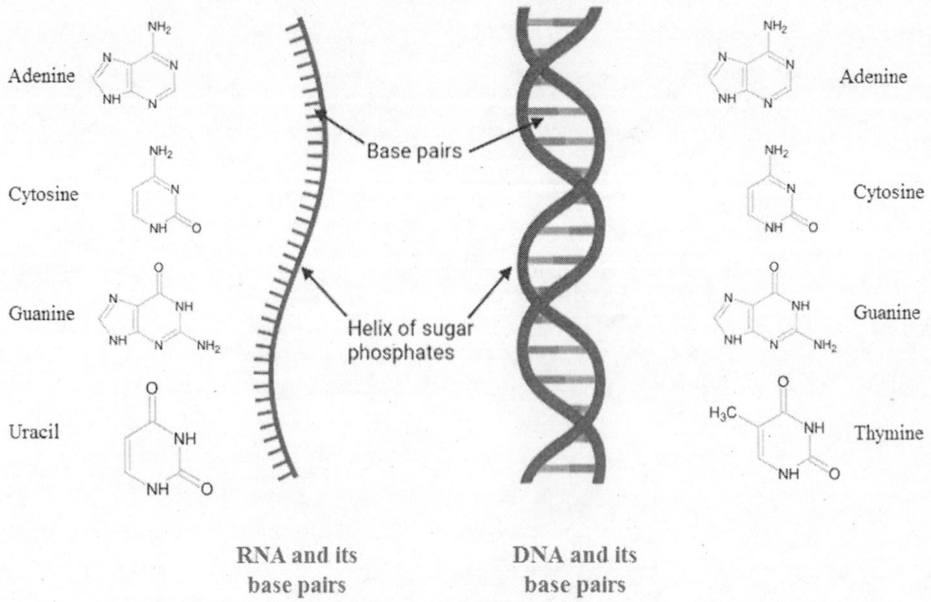

Fig. 4. 22 Nucleic acids

(Images created using BioRender® software)

4.5.1 DNA

All living things have genetic material called DNA. Eukaryotic cells have it in their nuclei, chloroplasts, and mitochondria. DNA is not contained within a nucleus in prokaryotes. The information that cells require to produce proteins is encoded by deoxyribonucleic acid. A DNA double helix consists of two right-handed spiral chains of a polynucleotide. Hydrophilic and hydrophobic interactions between the molecules that make up DNA and the water in a cell cause DNA to twist. The replication of DNA and the synthesis of proteins in our cells are significantly dependent on the double-helical structure of DNA. Both strands are antiparallel to each other which means that if one strand is in 3' to 5' direction then the other will be oriented in 5' to 3' direction. The strands are held together by hydrogen bonds between base pairs. Only specific kinds of base pairing take place. The only possible pairings for A and G are T and C. This is referred to as the base complementary rule. Or, to put it another way, the DNA strands are complementary to one another. If one strand has the sequence 5′-AATTGGCC-3′, the complementary strand would have the sequence 3′-TTAACCGG-5′. The base pairs are stabilized by forming two hydrogen bonds between adenine and thymine and three hydrogen bonds between cytosine and guanine. Hydrogen bonding is the key interaction that is responsible for the stabilization of DNA.

4.5.2 RNA

RNA, the other form of nucleic acid, is primarily involved in the creation of proteins. Messenger RNA (mRNA) serves as a bridge for DNA molecules to connect with other cell components. Other RNA types like rRNA, tRNA, and microRNA aid protein synthesis and its regulation. RNA is composed of nucleotides connected by phosphodiester bonds, just like DNA. The major differences between DNA and RNA are the bases and sugars present in them. In RNA, the nitrogenous base uracil (U) replaces thymine (T), and therefore Adenine pairs with Uracil in RNA. The sugar molecule present in RNA is ribose whereas the sugar present in DNA is Deoxyribose (Fig. 4.23).RNA typically has a single strand, unlike DNA. In contrast, most RNAs show internal base pairing between complementary sequences, creating a three-dimensional structure essential to their functionality. Messenger RNA (mRNA),ribosomal RNA (rRNA), transfer RNA (tRNA), and microRNA (miRNA) are the four main kinds of RNA.

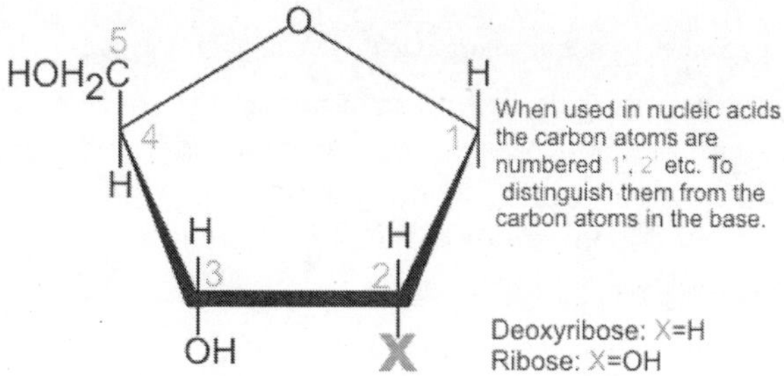

Fig. 4. 23 Difference in sugars present in DNA & RNA

Source: wikimedia.org (Creative Common License)

Messenger RNA (mRNA) - For the cell to interact with one another, mRNAs are created. The genetic instructions needed to build proteins are found in mRNA, obtained by decrypting DNA. Prokaryotic mRNA can begin synthesizing proteins without requiring structural modification (i.e., no cap on mRNA). A newly produced RNA transcript in eukaryotes is known as a pre-mRNA and must undergo maturation to become an mRNA. In eukaryotes, post-transcriptional processing involves capping the 5' end of mRNA with a guanosine triphosphate nucleotide. This helps mRNAs be recognized during translation or protein synthesis. To prevent enzymatic mRNA degradation, an mRNA's 3' end has several adenylate residues or a poly "A" tail attached to it. An mRNA's stability is influenced by both the 5' and 3' ends help maintain the stability of an mRNA

Transfer RNA (tRNA) - mRNA is translated into proteins by RNA molecules called tRNAs. Their circular structure consists of a 3' acceptor site, 5' terminal phosphate, D arm, T arm, and anticodon arm (Fig. 4.24). A tRNA's primary function is to carry amino acids to a ribosome complex on its 3' acceptor site with the help of aminoacyl-tRNA synthetase. To make proteins, the correct amino acid must be joined to a free tRNA by the enzymes known as aminoacyl-tRNA synthetases.

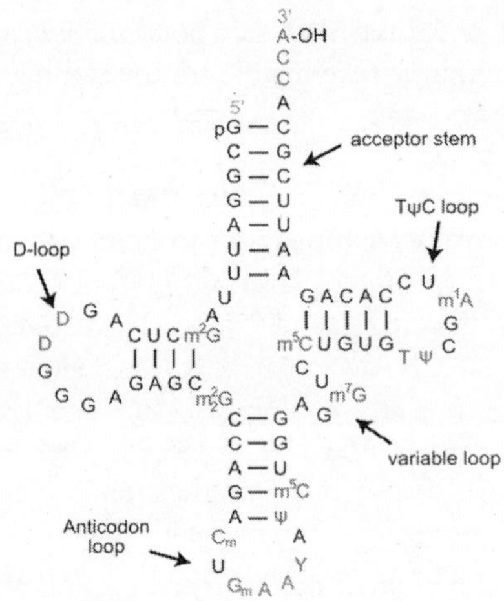

Fig. 4. 24 Transfer RNA (tRNA)

Source: wikimedia.org (Creative Common License)

A tRNA joined by an amino acid is known as an aminoacyl-tRNA. The kind of amino acid present in a tRNA is determined by the mRNA codon, a sequence of three nucleotides that codes for an amino acid. The mRNA codon's complementary anticodon, which specifies which amino acid to carry, is located on the anticodon arm of the tRNA. By removing cytochrome c from the body, tRNAs also control apoptosis.

Ribosomal RNA (rRNA) - Ribosomes are necessary for protein synthesis and are made up of rRNA and ribosomal proteins. There are two ribosomal subunits: a large and a small one. A prokaryotic 70S ribosome comprises the small 30S and large 50S ribosomal subunits. The 40S and 60S subunits combine to generate an 80S ribosome in eukaryotes. (Note: Here S is the svedberg unit corresponding to the sedimentation coefficient. It does not comply with the associative property of addition.) The ribosomes have three sites for binding aminoacyl-tRNAs and combining amino acids to form polypeptides: the exit (E), peptidyl (P), and acceptor (A) sites.

MicroRNA (miRNA) - Small, single-stranded, non-coding RNA molecules called miRNA attach to target mRNA to block the creation of proteins by one of two different processes. Primary miRNA (pri-miRNA), which joins the effector complex RNA-induced silencing complex (RISC), is cleaved twice to produce mature miRNA. The miRNA acts as a guide by base-pairing with the target mRNA to inhibit its production. In other words, microRNAs help regulate gene expression in cells.

4.5.3 Linkages in DNA & RNA

Phosphodiester Bonds - Phosphodiester bonds stabilize DNA and RNA. It is the phosphate group's bond with the sugar. The nucleotide polymers DNA and RNA are created by joining two distinct nucleotides via a diester bond (between phosphoric acid and two sugar molecules).

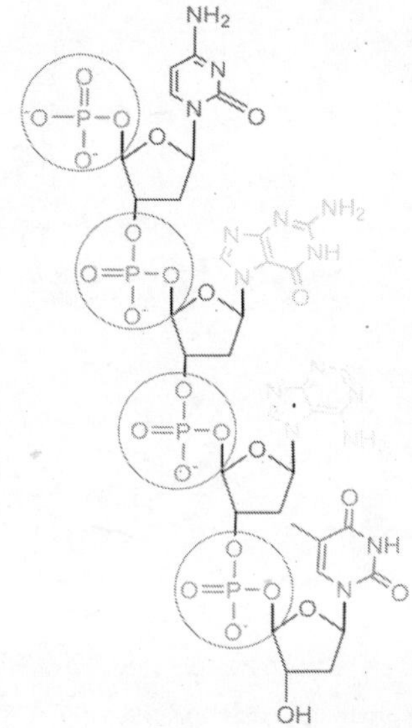

Fig. 4. 25 Phosphodiester bond in a strand of a polynucleotide

(Source: Created with ChemDraw software)

A phosphodiester bond is the link between the 3' carbon atom of one sugar molecule and the 5' carbon atom of another, such as ribose in RNA and deoxyribose in DNA. With two pentoses that have five carbons apiece and are connected by two ester bonds, the phosphate group produces strong covalent bonds.

Hydrogen Bonds - The positive hydrogen end of a polar N-H bond joins with a pair of electrons on either nitrogen or carbonyl oxygen to create a hydrogen bond. Another crucial characteristic of these "complementary" base pairs is that a purine base (adenine or guanidine) invariably links to a pyrimidine base (cytosine or thymine). This implies that the separation between the two strands is always controlled and the same (three rings and hydrogen bonds). A-T binds together in two hydrogen bonds, while the C-G pair makes three. The two DNA strands are held together by hydrogen bonds formed between complementary nucleotides. Chemical bonding does not apply to hydrogen bonds as they are susceptible to disruption. As a result, the DNA strands can split apart for replication and transcription of DNA to RNA (copying DNA to DNA).

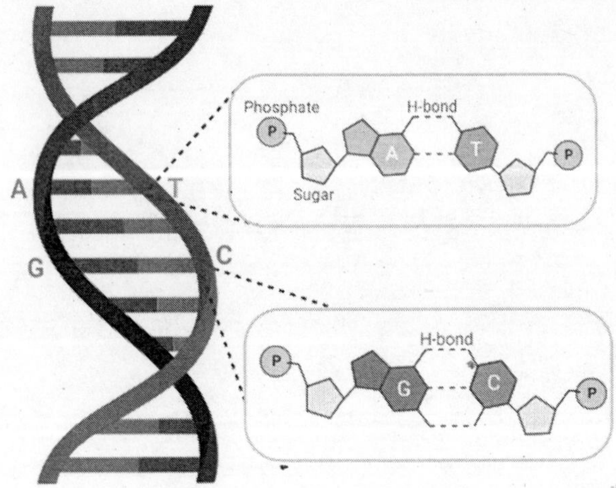

Fig. 4. 26 Hydrogen bonding in base pairs

(Images created using BioRender® software)

4.5.4 DNA packaging

Histones aid in the packaging of chromosomal DNA into tiny nuclei. These proteins have positive charges that firmly cling to DNA with negative charges to form structures known as nucleosomes.

Eight histone proteins and 1.65 times of DNA make up each nucleosome. A 30-nanometer chromatin fiber made of folded nucleosomes creates loops that are 300 nanometers in length on average. A 250 nm-wide fiber, snugly wound into a chromosome's chromatid, is built by compressing and folding 300 nm fibers.

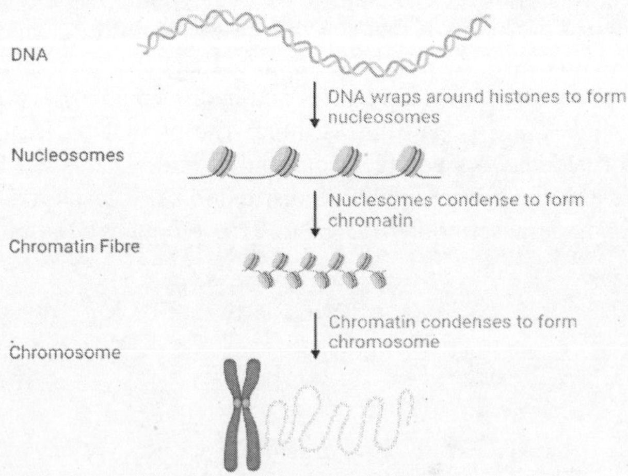

Fig. 4. 27 DNA packaging in eukaryote

(Images created using BioRender® software)

4.6 LIPIDS

Lipids are chemical substances connected to fatty acids, potentially utilized by living cells, and are mainly soluble in organic solvents (alcohol, ether, etc.). Lipids are not polymers, unlike proteins, polysaccharides, and nucleic acids. Simple, complex, and derived lipids are the three primary categories, and each of them is further subdivided into various groups.

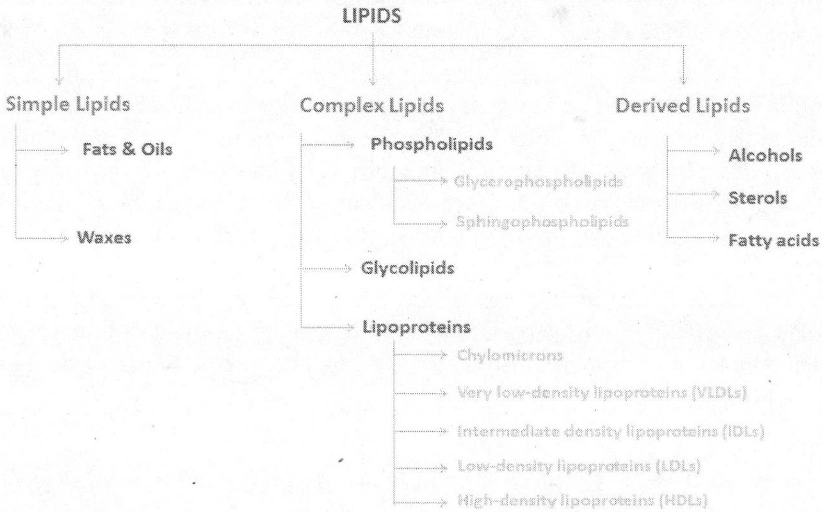

4.6.1 Classification of Lipids

Simple lipids - They are esters of fatty acids with polyols. These are mainly of two types:

Fats and oils (triacylglycerols) - These are fatty acid esters derived from glycerol. Triglycerides help absorb fat-soluble vitamins, store energy, and insulating cells. Hydrocarbon chains with varying lengths and saturation levels that finish in carboxylic acid groups are known as fatty acids. The only distinction between oil and fat is one of nature. At room temperature, oil is a liquid, whereas fats are solid.

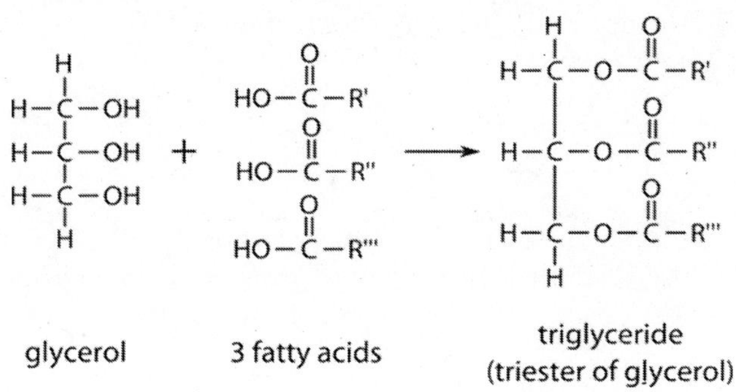

Fig. 4. 28 Formation of Triglyceride

(Source: Created with ChemDraw software)

Waxes - Esters of non-glycerol alcohols with fatty acids (often long-chain). They could be aliphatic or alicyclic alcohol. They offer protection, particularly to plants whose leaves are covered in wax. In humans, cerumen, sometimes referred to as earwax, aids in safeguarding the ear canal's skin. Candles, lubricants, cosmetics, ointments, polishes, and other products are all prepared using wax.

Complex (or compound) lipids - These are fatty acid esters having extra groups in the alcohols, such as phosphate, nitrogenous bases, carbohydrates, proteins, etc. Following is a broader analysis of them:

Phospholipids - These substances usually include a nitrogenous base and phosphoric acid. This is in addition to fatty acids and alcohol. The phosphate group is hydrophilic and fatty acid is hydrophobic. Phospholipids are hence amphipathic. The bilayer arrangement of phospholipids in the cell membrane protects the cell and acts as a barrier to some chemicals. These phospholipids are arranged into a lipid bilayer in the cell membrane (Fig. 4.27) with the hydrophilic phosphate group facing out and the hydrophobic tail buried into the bilayer. This arrangement limits polar molecules from freely diffusing through the membrane.Oxygen and water are examples of small, polar molecules efficiently diffusing into and out of the cell. Transport proteins are required to support large, polar molecules as glucose moves around because they cannot do it independently.

(i) Glycerophospholipids: These phospholipids, such as lecithin and cephalin, include glycerol as the alcohol.

(ii) Sphingophospholipids: The alcohol in this class of phospholipids, such as sphingomyelin, is called sphingosine.

Glycolipids - These lipids have a nitrogenous base, a fatty acid, and a carbohydrate. They are also known as glycosphingolipids since the alcohol in them is sphingosine. Some compounds, such as cerebrosides and gangliosides, lack phosphate and glycerol.

Lipoproteins - Lipid and protein macromolecular complexes.

Other complex lipids - Among the other complex lipids are sulfolipids, amino lipids, and lipopolysaccharides.

Derived lipids - These are the compounds produced by the hydrolysis of group 1 and group 2 lipids, which are lipid-like substances. In addition to glycerol and other alcohols, these substances include fatty acids, mono- and diacylglycerols, lipid-soluble vitamins, steroid hormones, hydrocarbons, and ketone bodies.

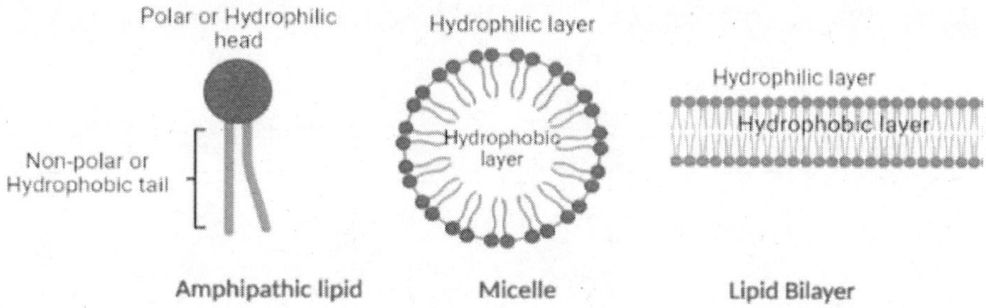

Fig. 4. 29 Arrangements of Lipid

(Images created using BioRender® software)

Amphipathic molecules generally represent substances that include key polar and hydrophilic constituents and hydrophobic molecules. Such compounds are essential in biology, as these play a critical role in forming the biological barrier.

Micelles are lipid monolayers with a fatty acid core, a polar surface, or a polar core with fatty acids on the surface (inverted micelle). Polar head groups, which typically make up the micelle's exterior surface, are present in micelles. They are polar; therefore, they face the water. The hydrophobic tails are non-polar and lie within the water. Micelle-derived fatty acids typically contain a single hydrocarbon chain rather than two hydrocarbon tails. As a result, there is less steric restriction for them to assume a spherical shape within a fatty acid.

On the other hand, fatty acids from glycolipids and phospholipids contain two hydrophobic chains. In an aqueous medium, lipid bilayers develop quickly and spontaneously and are maintained by hydrophobic contacts, attractive Van der Waals forces, and electrostatic interactions. In the living cell, the lipid bilayer's job is to provide a barrier between the two sides of the membrane.

4.6.2 Lipoproteins

A hydrophobic triacylglycerol and cholesterol esters core is encased in an amphipathic protein, phospholipid, and cholesterol shell to form the globular, micelle-like particles known as lipoproteins. The apolipoproteins (apoproteins) on the lipoproteins' surface assist in lipid solubilization and lipoprotein targeting to the appropriate organs. Chylomicrons, very low-density lipoproteins (VLDLs), intermediate-density lipoproteins (IDLs), low-density lipoproteins (LDLs), and high-density lipoproteins (HDL) are the five primary forms of lipoprotein that are categorized based on their physical and functional characteristics. Lipoproteins transport triacylglycerols, cholesterol, and phospholipids as they move throughout the body. Chylomicrons are synthesized in the intestine and transport dietary triacylglycerols to skeletal muscle and adipose tissue and dietary cholesterol to the liver. VLDLs are synthesized in the liver and transport triacylglycerols, cholesterol, and phospholipids to other tissues. HDLs are synthesized in the blood and extract cholesterol from cell membranes, converting it into cholesterol esters. Some of the cholesterol esters are then transferred to VLDLs.

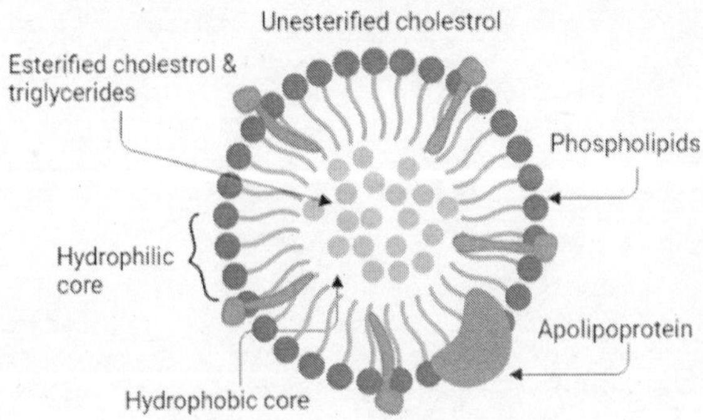

Fig. 4. 30 Structure of Lipoproteins

(Images created using BioRender® software)

In a nutshell, the chapter discusses the basics of biomolecules and how they play indispensable roles in all life processes. Biomolecules are crucial for maintaining the structure and effective functioning of living cells. The chapter deals with the function of these molecules, how they interact, and what reactions they undergo.

UNIT SUMMARY

Although living things come in an astonishing variety, their chemical makeup and metabolic processes are similar. There are many biomolecules with modest molecular weights (1000 Da). Some organic substances found in living creatures include amino acids, monosaccharide and disaccharide sugars, fatty acids, glycerol, nucleotides, nucleosides, and nitrogen bases. The four primary types of biomolecules are carbohydrates, lipids, proteins, and nucleic acids.

Major structural components of cells include proteins. They also operate as enzymes and catalysts for the bulk of chemical events that occur in living things, transporting nutrients and other substances into and out of cells.

Both plants and animals depend on carbohydrates for survival. They are produced by photosynthesis in plants. Energy-wise, carbohydrates are a good fuel source. Simple sugars include monosaccharides. They contain a free keto or aldehyde group. Two monosaccharides joined together by a glycosidic link form disaccharides. Carbohydrates polymer called polysaccharides hydrolyzes to produce more than ten monosaccharide units.

Lipids are the basic constituents of biological membranes and are often fatty acid esters. Lipid molecules are energy storage molecules because they can store a lot of energy. Lipids are divided into three categories based on their chemical components: simple lipids, complex lipids, and derived lipids. Simple lipids are long-chain alcohols or glycerol esters of fatty acids. Esters of fatty acids with alcohol are called compound lipids because they have additional groups in them. Simple and complex lipids are hydrolyzed to produce derived lipids.

Nucleotides are the building blocks of nucleic acids. A nitrogenous base, a pentose sugar, and a phosphate group make up a nucleotide. The genetic materials DNA and RNA are made up of polymerized nucleotides.

EXERCISES

Multiple Choice Questions

1) Carbohydrates are classified as monosaccharides, oligosaccharides, and polysaccharides based on their behavior on hydrolysis.
   A. True
   B. False

2) Polysaccharides are formed by _______
   A. Glycosidic linkages
   B. Peptide linkage
   C. Phospho-diester linkage
   D. Vanderwaal forces

3) Which among the following is incorrect about polysaccharides?
   A. Cellulose can't be digested by mammals unless they contain microbes that digest cellulose in their gut
   B. Polysaccharides mainly act as food storage or structural materials
   C. Starch comprises two forms, namely, amylose (water insoluble) and amylopectin (water soluble)
   D. Amylopectin has α-1, 4-linkage and 1, 6-glycosidic linkage

4) Primary storage of food in animals is through _______
   A. Fats
   B. Glucose
   C. Glycogen
   D. Galactose

5) Which among the following is incorrect about amino acids?
   A. There are 20 naturally occurring amino acids
   B. All the amino acids are chiral
   C. Amino acids are classified as acidic, basic, or neutral based on the relative number of amino and carboxylic groups
   D. Amino acids are classified as essential amino acids and non-essential based on whether they are synthesized in the body or not.

6) Non-essential amino acids are those amino acids that our body doesn't contain and acquire through food.
   A. True
   B. False

7) Proteins are polymers of amino acids that are connected by __________

A. Peptide linkage
B. Glycosidic linkage
C. Phosphodiester linkage
D. Vanderwaal linkages

8) Which among the following nucleotides is not present in DNA?

A. Adenine
B. Thymine
C. Cytosine
D. Uracil

9) Which among the following is incorrect about lipids?

A. These are heterogeneous organic compounds comprising C, H, and O
B. These are water insoluble and release more energy as compared to carbohydrates
C. These also act as water barriers and messengers and are constituents of plant
D. Phospholipids are an example of simple fatty acids

10) Which nutrient provides the maximum energy on the breakdown?

A. Carbohydrates
B. Fats
C. Fibres
D. Proteins

11) The main interaction responsible for plasma membrane stabilization:

A. Hydrophobic interactions
B. Hydrophilic interactions
C. Covalent bonds
D. Ionic bonds

12) The phospholipid contains non-polar _______ tails and contains fatty acid chains.

A. Hydrophobic
B. Hydrophilic
C. Both a & b
D. Amphiphilic

13) The phospholipid consists of

A. Polar head
B. Non-Polar tail
C. both a & b
D. Non-Polar head & Polar tail

14) What is the site of rRNA synthesis?
   A. Nucleolus
   B. Chromatin
   C. Perinuclear space
   D. Centrosomes

15) Fatty acids are amphipathic by nature.
   A. True
   B. False

16) Which of the following is not a component of a phospholipid?
   A. Phosphate
   B. Alcohol
   C. Glycerol
   D. Protein

17) A lipid is a polymer made up of
   A. Glucose molecules
   B. Amino acids
   C. Nucleotides
   D. Fatty acids and glycerol

18) If the DNA strand has a nitrogenous base sequence 5'-ATTGCC-3', then reading from 3' to 5' direction, state the sequence of bases that mRNA will have.
   A. TAACGG
   B. UGGACC
   C. UAACGG
   D. TACCGA

19) The bases are held together in a DNA double helix by hydrogen bonds. These bonds are
   A. Ionic bonds
   B. Covalent bonds
   C. Non-covalent bonds
   D. Van der Waals forces

20) On complete hydrolysis of starch, we get
   A. Glucose
   B. Fructose
   C. both a & b
   D. Sucrose

21) Which of the following disaccharides is formed from two identical monosaccharide units?
A. Maltose
B. Lactose
C. Sucrose
D. Fructose

22) Which of the following is not an example of a pyrimidine base?
A. Guanine
B. Uracil
C. Cytosine
D. Thymine

23) How many peptide bonds are present in a tripeptide?
A. 3
B. 1
C. 2
D. 4

24) Plant cell wall is made up of:
A. Cellulose
B. Starch
C. Sucrose
D. Glycogen

25) A nucleotide is formed of which of the following units?
A. nitrogen base and phosphate
B. nitrogen base, sugar and phosphate
C. nitrogen base and sugar
D. sugar and phosphate

26) When the negatively charged DNA combines with the positively charged histone octamer, which of the following is formed?
A. Nucleus
B. Nucleoid
C. Nucleosome
D. Nuclepores

27) Each nucleosome is made of DNA wrapped around ____ histone(s).
A. 1
B. 2

C. 8
D. 7

28) What are the thread-like stained structures present in the nucleus known as,?
A. Chromosome
B. Chromatid
C. Chromatin
D. Chloroplast

29) The highest phospholipids content is found in
A. Chylomicrons
B. VLDL
C. LDL
D. HDL

Answers: 1)A; 2)A; 3)C; 4)A; 5)B; 6)B; 7)A; 8)D; 10)D; 11)B; 12)A; 13)A; 14)C; 15)A; 16)A; 17)D; 18)D; 19)C; 20)C; 21)A; 22)A; 23)A; 24)C; 25)A; 26)B; 27)C; 28) C; 29)C; 30)D

Short Answer Questions

1) What type of linkages hold together monomers of DNA?
2) Give the differences between DNA & RNA
3) What are reducing and non–reducing sugars?
4) Classify the following into monosaccharides, oligosaccharides, and polysaccharides - i) Starch ii) fructose iii) sucrose iv) lactose iv) maltose
5) What are the functions of lipids in living organisms?
6) Define: 1) Peptide Linkage, 2) Enzymes 3) Glycosidic Linkage
7) What are essential and non-essential amino acids in human food? Give one example of each type.
8) How are micelles and lipid bilayers formed?

Long Answer Questions

1) Give the functions of lipids?
2) What are nucleic acids? Describe the structure of DNA.
3) What are the main functions of the cell wall?
4) What is the role of the plasma membrane in the compartmentalization of the cell?
5) How are Lipids classified?
6) Give the differences between Starch and Cellulose.

7) Explain the types of RNA
8) How is DNA packaged into chromosomes?

REFERENCES AND SUGGESTED READINGS

- Hames, D., & Hooper, N., (Third Edition). (2005). *Biochemistry*. Taylor & Francis Group.
- Nelson, D. L., & Cox, M. M., (Sixth Edition). (2013). *Lehninger Principles of Biochemistry*. Macmillan Learning.
- Satyanarayana, U., & Chakrapani, U., (Fourth Edition). (2013). *Biochemistry.* Elsevier.
- Jain, J. L., Jain, N., Jain, &Sunjay., (2005). *Fundamentals of Biochemistry*. S. Chand Publishing.
- Khalsa, N., (2008). *Essentials of Biochemistry.* Aavishkar Publishers.
- Wilson, K., & Walker, J., (2010). (Seventh Edition). *Principles and Techniques of Biochemistry and Molecular Biology*. Cambridge University Press.

Dynamic QR Code for Further Reading

5 Enzymology

UNIT SPECIFICS

Through this unit we have discussed the following aspects:

- *Classification of Enzymes*
- *Functions of Coenzyme and their Precursors*
- *Enzyme Kinetics*
- *Enzyme Inhibition*
- *Example of enzyme with applications*
- *RNA catalysis*

The ***practical applications of the topics are discussed*** *for generating further curiosity and creativity as well as improving problem solving capacity.*

Besides giving a large number ***of multiple choice questions as well as questions of short and long answer types*** *marked in two categories following lower and higher order of* ***Bloom's taxonomy****, assignments through a number of numerical problems, a list of references and suggested readings are given in the unit so that one can go through them for practice. It is important to note that for getting more information on various topics of interest some* ***QR codes*** *have been provided in different sections which can be scanned for relevant supportive knowledge.*

After the related practical, based on the content, there is a ***"Know More"*** *section. This section has been carefully designed so that the supplementary information provided in this part becomes beneficial for the users of the book. This section mainly highlights the initial activity, examples of some interesting facts, analogy, history of the development of the subject focusing the salient observations and finding, timelines starting from the development of the concerned topics up to the recent time, applications of the subject matter for our day-to-day real life or/and industrial applications on variety of aspects, case study related to environmental, sustainability, social and ethical issues whichever applicable, and finally inquisitiveness and curiosity topics of the unit.*

RATIONALE

This fundamental unit helps students to understand that without catalysis life would not have existed on earth. It explains on how to monitor enzyme catalyzed reactions. How does an enzyme catalyze reactions. Enzyme classification. Mechanism of enzyme action. It discusses at least two examples. Enzyme kinetics and kinetic- parameters and why should we know these parameters to understand biology?

PRE-REQUISITES

Biology: Cellular structure, Prokaryotic and Eukaryotic cell (Class XII)

Chemistry: Bond formation (Class XII)

UNIT OUTCOMES

List of outcomes of this unit is as follows:

U5-O1: Able to classify enzymes and coenzymes

U5-O2: Explain the functions and properties of coenzymes and cofactors

U5-O3: Realize the mechanism of enzyme action

U5-O4: Apply enzyme kinetics and its inhibition properties to solve biocatalysis problems

U5-O5: Discuss two examples of enzymes in biocatalysis reaction with its applications

Unit-5 Outcomes	EXPECTED MAPPING WITH COURSE OUTCOMES (1- Weak Correlation; 2- Medium correlation; 3- Strong Correlation)				
	CO-1	CO-2	CO-3	CO-4	CO-5
U5-O1	1	3	1	3	2
U5-O2	1	3	1	3	2
U5-O3	1	1	1	3	1
U5-O4	1	1	1	3	1
U5-O5	1	2	1	3	2

5.1 INTRODUCTION

One of the common fundamental characteristics of all living organisms is that thousands of chemical reactions occur in them at relatively low temperature. Moreover, these reactions happen at very high rates and in a highly regulated manner. Enzymes are the organic substances that act as biocatalyst and accelerate those biochemical reactions in living systems. Chemically all enzymes are proteins. Beadle (1948) defined enzymes as indispensable compounds that play a key role in metabolism by bringing direction and control to the physiological process of living cells. Any change in enzyme complement of living cell is immediately reflected in the change in the physiological and biochemical process of the cell. Thus, Enzymes play a very important role in the living world. Enzymology is a branch of biochemistry which deals with the study of enzymes, their kinetics, structure, and function, as well as their relation to each other.

5.2 ENZYME CLASSIFICATION

As more enzymes were discovered, chemists recognized the need for a more systematic and chemically informative identification scheme. In the current numbering and naming scheme, under the oversight of the Nomenclature Commission of the International Union of Biochemistry, enzymes are arranged into six groups according to the general type of reaction they catalyze (Table 5.1)

1), with subgroups and secondary subgroups that specify the reaction more precisely.

Table 5. 1 Classification of enzymes

adopted by I.U.B.

Class No.	Class type	Type of reaction
1	Oxidoreductases	Oxidation -reduction reaction $A^- + B \leftrightharpoons A + B^-$
2	Transferase	Transfer of groups A-B + C ⇆ A + B-C
3	Hydrolases	Hydrolytic reaction A-B + H_2O ⇆ A-H + B-OH
4	Lyases	Addition or removal of double bonds X Y | | A-B ⇆ A=B + X-Y
5	Isomerase	Transfer of group within molecule to yield isomeric form X Y Y X | | | | A-B ⇆ A-B

6	Ligases	Condensation of two molecule coupled through ATP hydrolysis A + B ⇆ A-B

5.3 ISOZYMES

Isozymes (also known as isoenzymes) are enzymes that differ in amino acid sequence but catalyze the same chemical reaction. These enzymes usually display different kinetic parameters (i.e. different K_M values), or different regulatory properties. Isozymes are usually the result of gene duplication, but can also arise from polyploidisation or hybridization. In terms of kinetics, isoenzymes have the capability to fine tune their enzymatic rate constants K_M and K_{cat}. This adaptation allows for the proper use of the enzyme based on its environment. A relevant example of such isozymes is lactate dehydrogenase (LDH). This enzyme is used to catalyze the synthesis of glucose in anaerobic metabolism of glucose. The isozymes of this enzyme are divided into two forms, the H isozyme and the M isozyme. The H isozyme is expressed more in the heart, whereas the M isozyme is expressed more frequently in the skeletal muscle. Another difference is that the H isozyme functions better in aerobic environments such as the heart, whereas the M isozyme functions better in anaerobic environments such as the muscle, where strenuous activity may deplete the oxygen supplies.An another prominent example of an isozyme is glucokinase, a variant of hexokinase which is not inhibited by glucose 6-phosphate. Its different regulatory features and lower affinity for glucose (compared to other hexokinases), allows it to serve different functions in cells of specific organs, such as control of insulin release-by-the-beta-cells-of-the-pancreas-or-initiation-of-glycogen-synthesis-by-liver-cells.Both-of-these-processes-must-only-occur-when-glucose-is-abundant,-or-problems-occur.

5.4 ENZYME ASSOCIATED METAL IONS AND COFACTORS

Many enzymes require cofactors for their catalytic activity. The cofactor may be a complex organic molecule called a coenzyme (Table 5.2) or a metal ion such as Fe^{2+}, Mn^{2+}, Zn^{2+}, or Mg^{2+} (Table 5.3). Holoenzymes are composed of an enzyme and a cofactor. In such cases, the protein component is the cofactor requiring an enzyme called apoenzyme Coenzymes take part in catalysis transiently and are carriers of specific functional groups. Most coenzymes are derived from vitamins (organic nutrients needed in small amounts in diet). When a coenzyme or metal ion is tightly bound through the covalent bond with the enzyme protein, it is called a **prosthetic group.**

Table 5. 2 Function of co enzyme and their precursors

Coenzyme	**Precursor vitamin**	**Role in the catalytic reaction**
Biocytin	Biotin (vitamin B7)	Transfer of CO2

Coenzyme B12 (5'-adenosylcobalamin)	Vitamin B12	Transfer of an alkyl group
Flavin adenine dinucleotide (FAD)	Riboflavin (vitamin B2)	Transfer of electrons
Coenzyme A	Pantothenic acid (vitamin B3)	Transfer of acyl and alkyl group
Nicotinamide adenine dinucleotide (NAD	Niacin (vitamin B5)	Transfer of hydride
Pyridoxal phosphate	Pyridoxine (vitamin B6)	Transfer of amino group
Thiamine pyrophosphate	Thiamine (vitamin B1)	Transfer of aldehydes
Tetrahydrofolate	Folic acid (vitamin B9	Transfer of one carbon group

Table 5. 3 Examples for metal ions as a cofactors

Metal Ions	**Enzyme name**
Fe^{2+} or Fe^{3+}	Catalase, peroxidase, cytochrome oxidase
Cu^{2+}	Cytochrome oxidase
Mg^{2+}	DNA polymerase
Mn^{2+}	Arginase
K^{+}	Pyruvate kinase
Mo^{2+}	Nitrogenase, nitrate reductase
Zn^{2+}	Carbonic anhydrase, alcohol dehydrogenase
Ni^{2+}	Urease

5.5 MECHANISM OF ACTION OF ENZYMES

The enzymes act on a substrate by combining with the substrate molecules to form an enzyme-substrate complex. Enzymes (E) have specific active sites for the attachments of substrate (S) molecule where an enzyme can form intimate relationship with the substrate. It is presumed that the enzyme-substrate combination (ES) brings about a deformation in some of the bonds in substrate molecule which favours the reaction to produce a product.

$$E + S \rightleftarrows ES \rightleftarrows EP \rightleftarrows E + P$$

The binding of enzyme and substrate is highly specific. A given enzyme usually binds to only one kind of substrate. The specificity of enzyme-substrate interaction arises mainly from hydrogen bonding and shape of the active site (which rejects molecules that do not have sufficiently complementary shape).

There are two theories explaining the enzyme-substrate interaction.

5.5.1 Lock-And-Key Theory

This theory was proposed by Fisher (1894). According to this theory, the formation of enzymes-complex during enzyme action is analogous to the fitting of lock and key. It is believed that the enzyme and the substrate both have strictly complementary structures which during complex formation fit to each other like a specific key in a particular lock. Here the enzyme is like a lock and substrate is like a key (Fig. 5.1). Only a specific key (substrate) with correctly positioned teeth fits into the keyhole (active site) of the lock (enzyme). Thus, the active site of the enzyme is a rigid and pre-shaped template where only a specific substrate can bind. In simple words as a particular lock can be opened by a particular key, in a same way particular enzyme acts on a particular substrate. This theory gave a satisfactory explanation for enzyme action, however, this theory fails to explain the flexible nature of an enzyme.

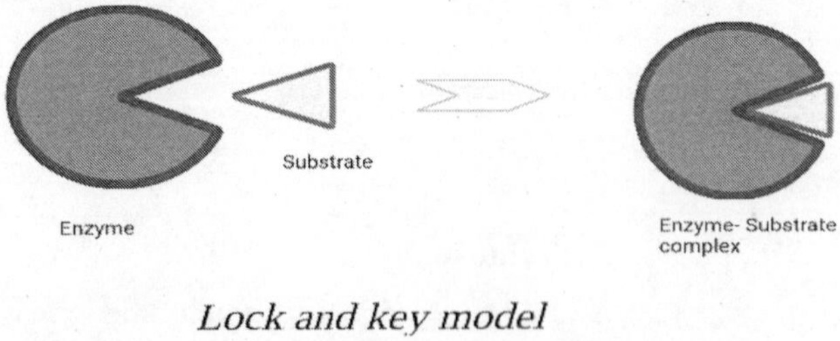

Fig. 5. 1 Schematic diagram of lock and key model

(Images created using BioRender® software)

5.5.1 Induced-fit theory

D. Koshland proposed this model in 1966. According to this model, the enzyme and substrate do not have strictly complementary structures, but the enzyme has a flexible active site with changes according to substrate configuration. Thus, theenzyme shows an induced fit mechanism during enzyme-substrate complex formation (Fig. 5.2). This model can be compared with hand and glove. The same glove can fit in the hands of many people. In other words, the glove adjusts itself in the hands of those who wear it. Similarly, the substrate induces conformational changes in the enzyme during enzyme action.

Fig. 5. 2 Schematic Diagram of Induced fit model

(Images created using BioRender® software)

5.8 MECHANISM OF ENZYME ACTION

All chemical reactions require some energy input to begin. The amount of energy needed for a reactant to undergo a chemical reaction is called activation energy. Energy is needed to break existing bonds before new bonds can be formed. The formation of new bonds may release more energy than was needed to break the original bonds. Even though there may be a net release of energy, the need for activation energy can act as a barrier to the chemical reaction occurring.

Enzymes lower the barriers that normally prevent chemical reactions from occurring by decreasing the required activation energy. The initial rise in energy is the energy input needed before the reaction will occur (activation energy). The subsequent drop in energy is the energy released by the reaction (Fig. 5.3). From the graph given below it is clear that the reaction requires less activation energy when an enzyme is present. This is why the addition of an enzyme allows a reaction to proceed at a much faster rate.

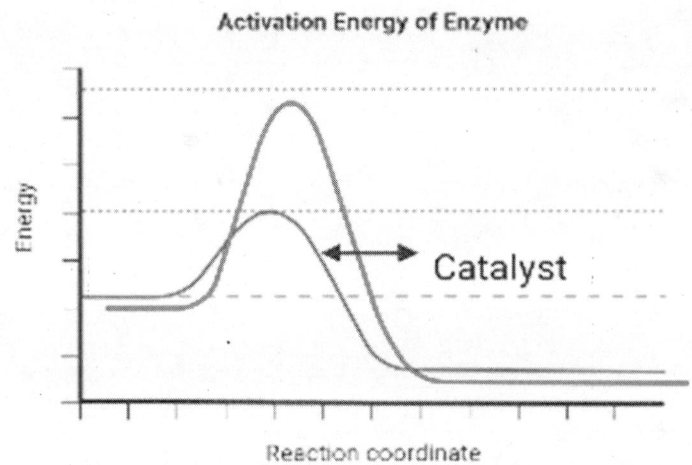

Fig. 5. 3 Enzymatic reaction with respect to Substrate consumption and time

5.9 KINETICS OF ENZYME-CATALYSED REACTION

During catalysis, substrate S binds at the active site of the enzyme E and results in formation of enzyme substrate complex ES, which is finally converted into product P. The reaction can be represented as:

$$E + S \rightleftarrows ES \rightleftarrows ES^* \rightleftarrows EP \rightleftarrows E + P$$

Where E forms weakly bonded complex ES with the substrate S. The ES complex decomposes to yield product P and the free enzyme E. Leonor Michaelis and Maud Menten in 1913 explained the kinetics of enzyme-catalyzed reactions. The most remarkable feature of this kinetics is that a specific ES complex is an intermediate during catalysis. Michaelis-Menten's theory of enzyme kinetics is the simplest one that accounts for the kinetic properties of many enzymes. Further simplifying the reaction, Michaelis-Menten derived the following equation for one substrate reaction.

$$Vo = \frac{Vmax[S]}{Km + [S]}$$

Where

Vo = Initial velocity

S= Substrate concentration (molar)

Vmax = maximum Velocity

Km = Substrate concentration at half of Vmax

The equation is called Michaelis-Menten equation. Where, K_M is called Michaelis constant, v0 is initial velocity, Vmax is maximum velocity of reaction, and [S] is substrate concentration. A graph of v0 against [S] results in rectangular hyperbola. Vmax is the maximum velocity at particular enzyme concentration. Vmax and K_M can be determined from the graph as shown in Fig. 5.4.

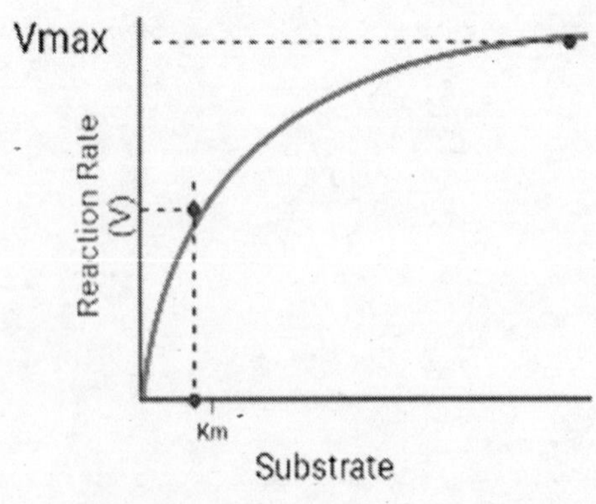

Fig. 5. 4 Difference between Vmax and Km

In the graph, we can see that at very low substrate concentration (when $[S] << K_M$), $v0=(Vmax/K_M)/[S]$, i.e., the reaction rate is directly proportional to substrate concentration. At high substrate concentration (when $[S]>>K_M$), v0 = Vmax, i.e., the reaction rate is maximum and independent of substrate concentration. When $[S]=K_M$, then v0= Vmax/2. Thus, K_M is the substrate concentration at which half of the maximum reaction rate is obtained. The maximum velocity, Vmax, represents the turnover number of an enzyme. Turnover number is the number of substrate molecules converted into the product by an enzyme molecule in a unit of time when the enzyme is fully saturated with substrate. It is equal to the kinetic constant k2, which is also called K_{cat}. When there is a substrate inhibition or activation due to binding of a second substrate molecule, the Michaelis-Menten Equation does not hold good. The study state and rapid equilibrium kinetics do not detailed information on existence of multiple intermediate or their life times.

5.10 LYSOZYME

Lysozyme is an enzyme in animal and human lacrimal gland secretions (or tears), gastric secretions, nasal mucus, and egg white. It was discovered in 1921 by Sir Alexander Fleming. The lysozymes catalyze the breakdown of certain carbohydrates found in the cell walls of certain bacteria (for example, cocci). As a result, in the case of lacrimal fluid, it protects the cornea of the eye from infection. Lysozyme is most effective against Gram-positive bacteria.

5.11.1 Structure of lysozyme

A single peptide chain of around 129 amino acids makes up the condensed structure of lysozyme from hen egg white (hen egg white lysozyme or egg lysozyme). The residues of amino acid are numbered from the terminal α-group (N) to the terminal carboxyl-group (C). The circles show every fifth and every tenth residue is numbered. And, the broken lines indicate the four disulfide bridges. In the ranges of 25 to 35, 90 to 100, and 120 to 125, alpha-helices can be seen. T7 lysozyme or Bacteriophage and T4 are some of the examples of lysozyme. The lysozyme enzyme breaks glycosidic bonds in peptidoglycans by hydrolyzing, attacking, and breaking them. Between the N-acetylmuramic acid (NAM) and the N-acetylglucosamine (NAG) fourth carbon atom, it attacks the peptidoglycans (which are found in the cell walls of bacteria, especially Gram-positive bacteria).

5.11.2 Applications

Lysozyme monomer exhibits potent antibacterial activity against Gram-positive organisms. This phenomenon has found a practical application in the food processing, medicine, and pharmaceutical industries. The use of lysozyme in the food processing industry is connected primarily with its application as a natural preservative. The enzyme is widely used as a preservative for meat, fish and their products, milk and dairy products, and fruit and vegetables. The pharmaceutical industry uses this enzyme to manufacture adjuvant drugs for antibiotics and analgesics in viral and bacterial infections to treat leukaemia and neoplastic diseases. Lysozyme is also used as a diagnostic agent, an indicator of the occurrence and the progression of pathological changes in humans and animals. The range of the practical applications of lysozyme may be considerably extended due to its modification. The enzyme, after modification exhibits a new specific activity to Gram-negative bacteria, resulting from dimerization, with no loss of activity against Gram-positive bacteria, characteristic of the monomer, as indicated in research studies. The dimeric form of lysozyme has been used to treat bacterial and viral animal diseases. A drug produced based on lysozyme dimer shows immune-stimulating and immune-corrective activity.

5.12 CHYMOTRYPSIN

Chymotrypsin is an enzyme used for digesting proteins. Chymotrypsin is found in the duodenum and selectively cleaves (cuts) off pieces of amino acids from the protein chain. Specifically, chymotrypsin cleaves phenylalanine, tyrosine, and tryptophan bonds, or aromatic amino acids. It cleaves these amino acids starting from the C-terminus of the protein. The primary structure shows that disulfide bonds have a crucial role in protein folding. The protein is spherical and consists of three polypeptide chains.

5.13 ENZYME INHIBITION

Substances that decrease the rate of an enzyme-catalyzed reaction are called enzyme inhibitors, and the process is known as enzyme inhibition. Enzyme inhibition can be classified as reversible inhibition and irreversible inhibition. In irreversible inhibition, the inhibitor binds tightly to the enzyme and does not dissociate. For example, the antibiotic penicillin acts as an inhibitor and binds with the enzyme transpeptidase, which synthesizes the bacterial cell wall. Hence, binding this drug to the enzyme prevents cell wall synthesis, thus killing the bacteria. In the same way, the drug aspirin inhibits the enzyme cyclooxygenase, thus reducing inflammation. In reversible inhibition, the inhibitor rapidly dissociates from the enzyme-inhibitor complex. There are three types of reversible inhibitions: competitive, non-competitive, and uncompetitive.

5.13.1 Competitive inhibition

In competitive inhibition, there is close resemblance in the structure of inhibitor I and substrate S, therefore, they both compete for the same active site on the enzyme. The enzyme can form enzyme-substrate ES complex or it can form enzyme-inhibitor EI complex (Figure 5.5) but not both ESI. Competitive inhibitors decrease the rate of reaction by reducing the amount of active enzyme molecules bound to a substrate.

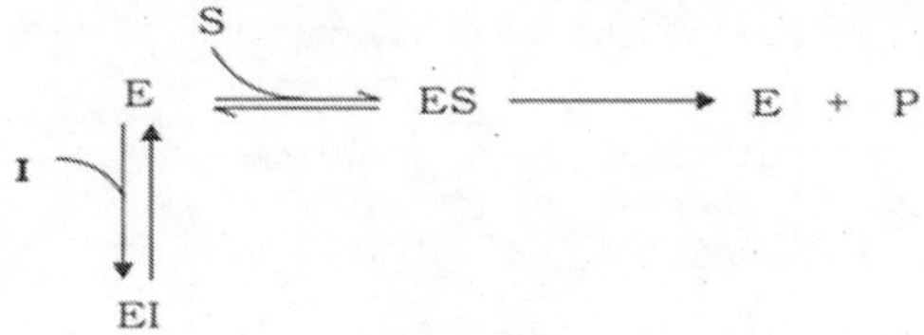

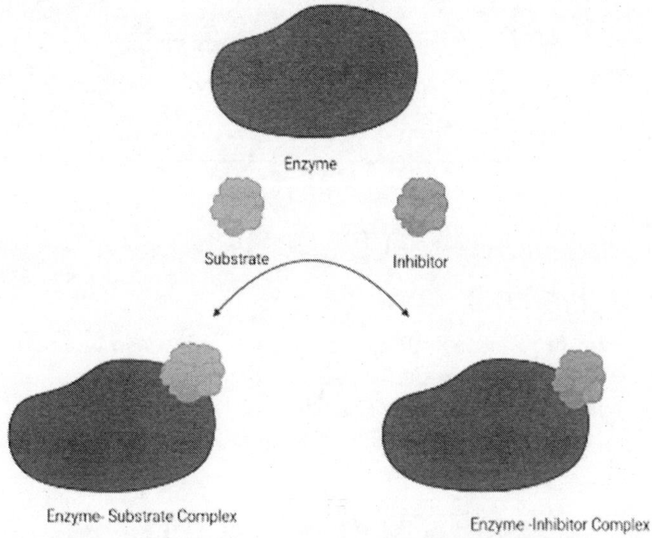

Fig. 5. 5 Mechanism of Competitive inhibition

(Images created using BioRender® software)

At very high substrate concentration, the chances of binding of inhibitor molecule to the enzyme will be reduced, so Vmax for the reaction will not be changed. However, the K_M which is substrate concentration at which v0=½ Vmax, is increased in presence of inhibitor and is denoted by symbol K'_M (Figure 5.6).

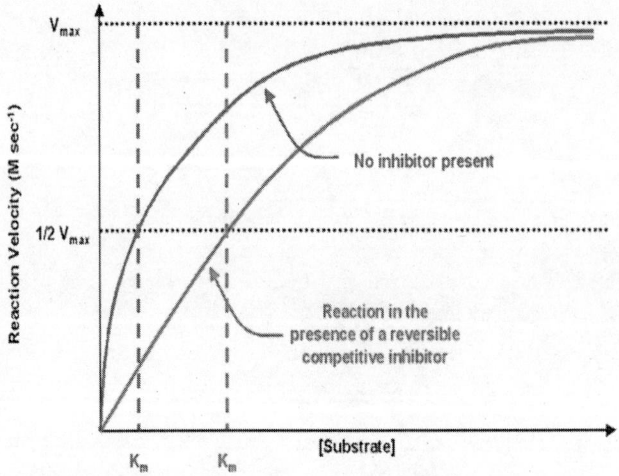

Fig. 5. 6 Effect of Competitive inhibition on Enzyme activity

5.13.2 Non-Competitive inhibition

In this type of inhibition, inhibitor has no structural similarity with substrate and binds with the enzyme at different site other than the active site. Therefore, there is no competition between S and I, and formation of ES, EI and ESI takes place. The inhibitor I and substrate S can bind simultaneously to the same enzyme molecule as their binding sites are different and hence do not overlap (Figure 5.7).

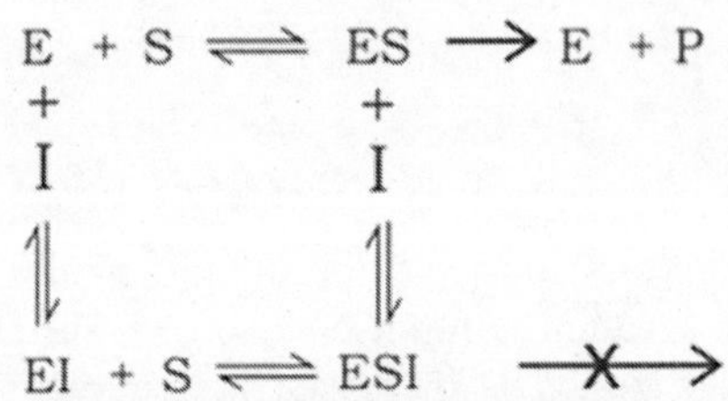

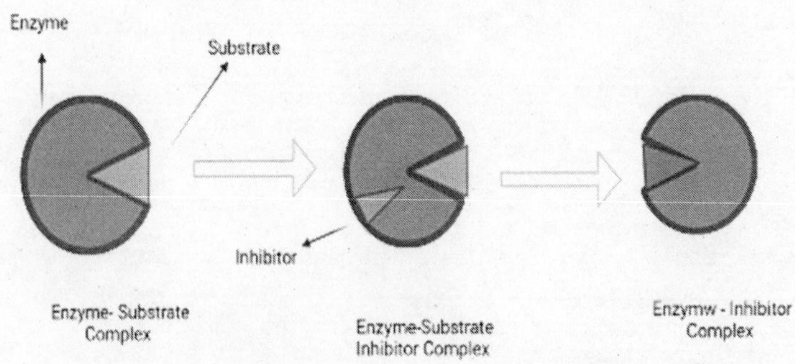

Fig. 5. 7 Mechanism of Non-Competitive inhibition

(Images created using BioRender® software)

Non-competitive inhibitor lowers the Vmax rather than decreasing the proportion of enzyme molecules bound to the S. Thus, the non-competitive inhibition, in contrast to competitive inhibition, cannot be overcome by increasing substrate concentration. The substrate can still bind to the EI complex. However, the ESI does not form the product. The I effectively lowers the active enzyme concentration and hence reduces the Vmax. There is no effect on K_M as the inhibitor decrease the amount of functional enzyme (Figure 5.8).

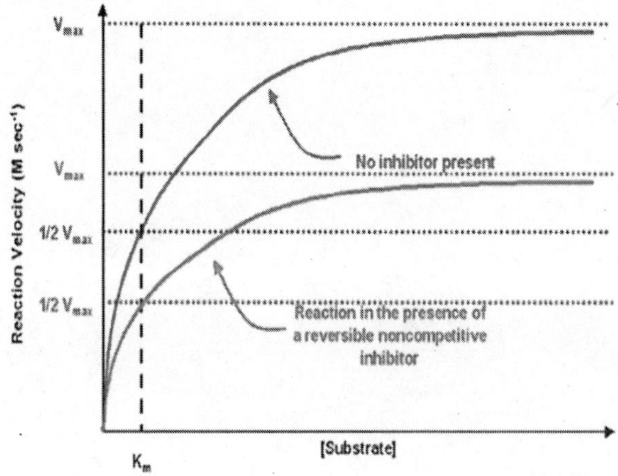

Fig. 5. 8 Effect of Non-Competitive inhibition on Enzyme activity

5.13.3 Uncompetitive inhibition

In this type of inhibition, the inhibitor does not bind to the free enzyme. It binds only to the enzyme-substrate (ES) complex directly, or its binding is facilitated by the conformational change after the substrate binds to an enzyme (Figure 5.9).

$$E + S \rightleftharpoons ES \rightleftharpoons E + P$$
$$ES + I \rightleftharpoons ESI \not\rightarrow$$

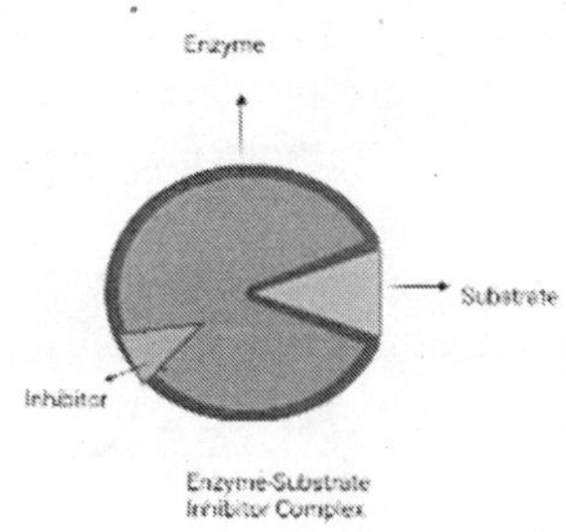

Fig. 5. 9 Mechanism of Uncompetitive inhibition

(Images created using BioRender® software)

In both cases, the inhibitor does not compete with the substrate for the same binding site. Therefore, the inhibition cannot be overcome by increasing substrate concentration. Both K_M and Vmax values are altered (Fig. 5.10).

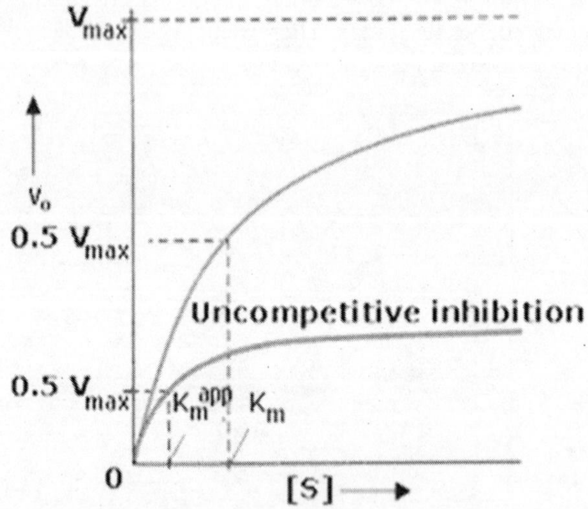

Fig. 5. 10 Effect of Uncompetitive inhibition on Enzyme activity

5.14 ALLOSTERIC ENZYMES

Allosteric enzymes are enzymes that have an additional binding site for effector molecules other than the active site. The binding brings about conformational changes, thereby changing its catalytic properties. There are two types of allosteric regulation based on substrate and effector molecules and based on the action performed by the regulator. Allosteric enzymes do not obey Michaelis-Menten kinetics. Allosteric enzymes result in a sigmoidal graph instead of a rectangular hyperbola when v0 is plotted against substrate concentration [S] (Fig. 5.11). Each subunit of allosteric enzymes also contains a regulatory site along with the active site. Regulatory molecules may reversibly bind to the regulatory site and alter the affinity of the enzyme for substrate binding. There are two models proposed for the mechanism of regulation of allosteric enzymes: Simple Sequential Model- Was given by Koshland. In this model, the binding of substrate induces a change in the conformation of the enzyme from T (tensed) to R (relaxed). Concerted or Symmetry Model-According to this model, there is a simultaneous change in all the subunits of an enzyme. All the subunits are either present in R form (active form) or T form (inactive form), having less affinity to a substrate. Many allosteric enzymes take part in the biochemical pathways so that the system is well controlled and modulated such as Aspartate Transcarbamoylase (ATCase), Glucokinase, and Acetyl-CoA Carboxylase.

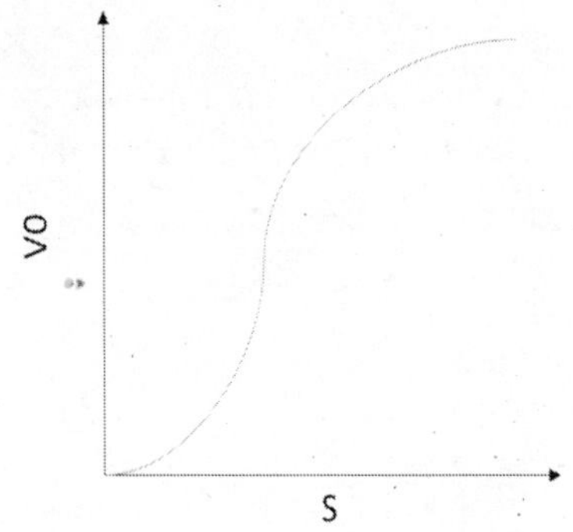

Fig. 5. 11 Effect of allosteric enzyme on the graph
(Images created using BioRender® software)

5.15 RNA CATALYSIS

Ribozymes (ribonucleic acid enzymes) are RNA molecules capable of catalyzing specific biochemical reactions, similar to the action of protein enzymes. The 1982 discovery of ribozymes demonstrated that RNA could be genetic material (like DNA) and a biological catalyst (like protein enzymes). It contributed to the RNA world hypothesis, which suggests that RNA may have been essential in the evolution of prebiotic self-replicating systems. The most common activities of natural or in vitro-evolved ribozymes are the cleavage or ligation of RNA and DNA and peptide bond formation. Within the ribosome, ribozymes function as part of the large subunit ribosomal RNA to link amino acids during protein synthesis. They also participate in various RNA processing reactions, including RNA splicing, viral replication, and transfer RNA biosynthesis. Examples of ribozymes include the hammerhead ribozyme, the VS ribozyme, Leadzyme, and the hairpin ribozyme.

Investigators studying the origin of life have produced ribozymes in the laboratory capable of catalyzing their own synthesis from activated monomers under very specific conditions, such as an RNA polymerase ribozyme.

5.16 Applications

Ribozymes have been proposed and developed for the treatment of disease through gene therapy. A type of synthetic ribozyme directed against HIV RNA called gene shears has been developed and has entered clinical testing for HIV infection. Similarly, ribozymes have been designed to target the hepatitis C virus RNA, SARS coronavirus, Adenovirus, and influenza A and B virus RNA. The ribozyme can cleave the conserved regions of the virus's genome, which has been shown to reduce the virus in mammalian cell culture.

CONCLUSION

At the end of this chapter, you learn about the function of enzymes in your body and daily life. You can comprehend how the kinetics of the enzymes contribute to the preservation of the environment, human health, and numerous bioprocesses.

UNIT SUMMARY

Life cannot exist without Enzyme. Enzymes are required in every step of metabolic reaction and play direct or indirect roles in all life functions, like cell division, digestion, brain function, production of secondary metabolite, etc.

EXERCISES

Multiple Choice Questions

1) Statement 1 Enzymes are very specific

   Statement 2 enzymes exhibit a enormous catalytic power

   Statement 3 enzymes are biocatalyst

   Statement 4 enzymes increase the rate of chemical reaction

   A. All statement are wrong
   B. All statements are correct except statement 2
   C. Statement 1,3 & 2 are correct
   D. All statement are correct

2) Statement- 1 All proteins are enzyme but all enzyme are not protein

   Statement -2 Many Enzymes activity are stop during the fever

   A. Both statements are correct
   B. both statements are wrong
   C. statement 1 is correct
   D. Statement 2 is correct

3) Enzyme + cofactor known as

   A. Apoprotein
   B. Apoenzyme
   C. Holoenzyme
   D. Coenzyme

4) In Enzyme classification EC stands for?

   A. Enzyme classification
   B. Enzyme commission
   C. Enzyme class
   D. Enzyme criteria

5.) Who proposed the lock and key model?

   A. Koshland

B. Myher
C. Emil fischer
D. Maud menten

6.) According to which model enzyme changes its shape when substrate is bind
A. Lock and key model
B. Induced fit model
C. Enzyme kinetics model
D. None

7.) Concentration of substrate at which reaction velocity reaches half its maximum velocity is called
A. Vmax
B. Vo
C. Km
D. Pka

Answer: 1) D; 2) B; 3) C; 4) B; 5) C; 6) B; 7) C

Short and Long Answer

1) How does a non-competitive inhibitor affect enzyme action?
2) Will an enzyme bind to its product? Explain.
3) There is no life without enzymes. Justify the statement.
4) How does the Michaelis-Menten equation explain why the rate of an enzyme-catalyzed reaction is proportional to the amount of enzyme?
5) What is the chemical basis of enzyme catalysis?

NUMERICAL PROBLEMS

1) An enzyme hydrolyzed urea at [S]= 0.003mmol/L with K_M value of 0.06 mmol/L. the initial velocity was 1.5 x 10^{-3} mmol/L and maximum velocity was 4.5 x 10^{-3} mmol/L.mol^{-1}. Calculate the K_M.
2) Urease enzyme hydrolyzed urea at [S]= 0.03 mmol/L with a Km value of around 0.06 mmol/L. The initial velocity observed was 1.5x10^{-3} mmol/L.min^{-1}. Calculate the-maximum velocity of-the enzymatic-reaction
3) An enzyme hydrolyzed a substrate concentration of 0.03 mmol/L, the initial velocity was 1.5x10^{-3} mmol/L.min^{-1} and the maximum velocity was 4.5 x 10^{-3} mmol/L.min^{-1}. Calculate the substrate concentration that gives a velocity of 3 x 10^{-3} mmol/L.min^{-1}.

PRACTICAL

Experiment to demonstrate that heart destroys activity of enzyme but not that of a catalyst.

KNOW MORE

Enzymes Assist Brain Function - The hypothalamus is the portion of our brain that regulates our endocrine system, and our emotional reactions and harmony play a significant role in this. The hypothalamus requires glucose to function properly and this glucose supplied from the liver is protein-dependent. The fact about enzymes is that it facilitates the activity of the liver and assists its glucose production activities. The carbohydrates are transported to different areas of the body by the enzymes in the blood to supply the muscles with the energy to function properly. Inadequate protein is transported in the blood with insufficient enzyme levels resulting in tiredness and lassitude found in the bodies particularly the brain.

REFERENCES AND SUGGESTED READINGS

- Principles of Biochemistry - David L. Nelson & Michael M.COX
- Essentials of Biochemistry - Pankaja Naik

Dynamic QR Code for Further Reading

6 Information Transfer

UNIT SPECIFICS

Through this unit we have discussed the following aspects:

- *Identification of DNA as a genetic material*
- *Double helical structure of DNA*
- *Hierarchy of DNA*
- *DNA replication*
- *Transcription: DNA to RNA*
- *Genetic code*
- *Protein Synthesis (Translation): RNA to protein*
- *Gene expression and regulation*
- *Lac operon Model*

The ***practical applications of the topics are discussed*** *for generating further curiosity and creativity as well as improving problem solving capacity.*

Besides giving a large number ***of multiple choice questions as well as questions of short and long answer types*** *marked in two categories following lower and higher order of* ***Bloom's taxonomy****, assignments through a number of numerical problems, a list of references and suggested readings are given in the unit so that one can go through them for practice. It is important to note that for getting more information on various topics of interest some* ***QR codes*** *have been provided in different sections which can be scanned for relevant supportive knowledge.*

After the related practical, based on the content, there is a ***"Know More"*** *section. This section has been carefully designed so that the supplementary information provided in this part becomes beneficial for the users of the book. This section mainly highlights the initial activity, examples of some interesting facts, analogy, history of the development of the subject focusing the salient observations and finding, timelines starting from the development of the concerned topics up to the recent time, applications of the subject matter for our day-to-day real life or/and industrial applications on variety of aspects, case study related to environmental, sustainability, social and ethical issues whichever applicable, and finally inquisitiveness and curiosity topics of the unit.*

RATIONALE

To understand how genetic information is transferred in a biological system. Understanding the mechanism through which our genetic material, DNA, duplicates itself for growth and cell division—revealing how the information stored in DNA is decoded to synthesize proteins to carry out all the cellular functions. This chapter will help you to understand the machinery working inside the cells.

PRE-REQUISITES

Biology: Cellular structure, Prokaryotic and Eukaryotic cell (Class XII)

Chemistry: Bond formation (Class XII)

UNIT OUTCOMES

List of outcomes of this unit is as follows:

U6-O1: Structure of genetic material

U6-O2: Synthesis of DNA

U6-O3: Role of hereditary material in the genetic transfer

U6-O4: Understanding the process of information transfer from DNA to RNA to protein

U6-O5: Understanding gene expression the classic model of Lac operon Model

Unit-6 Outcomes	***EXPECTED MAPPING WITH COURSE OUTCOMES*** *(1- Weak Correlation; 2- Medium correlation; 3- Strong Correlation)*				
	CO-1	**CO-2**	**CO-3**	**CO-4**	**CO-5**
U6-O1	1	2	3	1	1
U6-O2	1	2	3	1	1
U6-O3	1	2	3	1	1
U6-O4	1	2	3	1	1
U6-O5	1	2	3	1	1

6.1 INTRODUCTION

The genetic material of an organism contains all the information required to build and maintain that organism. It was first isolated by Freidrich Miescher in 1869 and was called nuclein as it had acidic properties. This was later called nucleic acid, with an elemental composition of carbon, hydrogen, oxygen, nitrogen, and phosphorus. Nucleic acids are the hereditary determinants of living organisms that pass information from one cell to a daughter cell at cell division and from one generation of an organism to the next through organisms' reproductive cells. Genetic material serves three crucial functions:

1. Genotypic function: It transmits genetic information from one generation to another.
2. Phenotypic function: It directs growth and differentiation. It also leads to the development of offspring into reproductive traits.
3. Evolutionary function: Mutations enable organisms to respond to environmental changes.

At the beginning of the 20th century, questions started to emerge on the nature of hereditary information being transferred from one parent to the offspring. In this search, proteins emerged as one of the major elements that perform most cellular functions and serve as building blocks for cellular structures that regulate gene expression and communication. The functions of a cell are primarily defined by its ability to produce proteins. Consequently, it was difficult to envision what kind of instructions the genetic material could contain. This led to a detailed investigation of the nature of genetic information, which led to the identification of DNA as the carrier of genetic information, the discovery of its structure, and finally, how it replicated, transformed, and transferred across generations.

6.1.1 DNA as a Genetic Material

Genetic material specifies the biological development of all life forms. DNA stores genetic information; it encodes the sequence of amino acid residues in proteins using the genetic code, triplet code. The role of DNA as genetic material was not discovered until 75 years after its discovery. In 1903, W. S. Sutton observed that the inheritance patterns of genes parallel the behavior of chromosomes during cell division, which led to the chromosome theory, the proposal that genes are located in chromosomes. Examination of cells by cytochemistry, which uses stains that bind specifically to just one type of biochemical, showed that chromosomes are made of DNA and protein in roughly equal amounts. But in the early 20th century, it was thought that all DNA molecules are the same. At the same time, proteins are highly variable molecules made up of 20 chemically distinct amino acids, leading to the misconception that genes are made up of proteins. The errors in understanding DNA structure lingered on, but by the late 1930s, it had become accepted that DNA, like protein, has immense variability. The notion that protein was the genetic material initially remained strong but was eventually overturned by the results of two important scientific experiments: A) Griffith's Experiment and B) Hershey Chase experiment.

6.1.2 Transforming principle: Griffith's Experiment

In 1928, Frederick Griffith conducted an experiment specifying the capability of bacteria to transfer genetic information. He studied the difference between pathogenic and non-pathogenic strains of pneumonia-causing bacteria, *Streptococcus pneumonia*. The pathogenic strain (Smooth- S strain) was surrounded by a capsule and caused pneumonia, whereas the non-pathogenic strain (Rough-R strain) did not have a capsule and did not cause pneumonia.

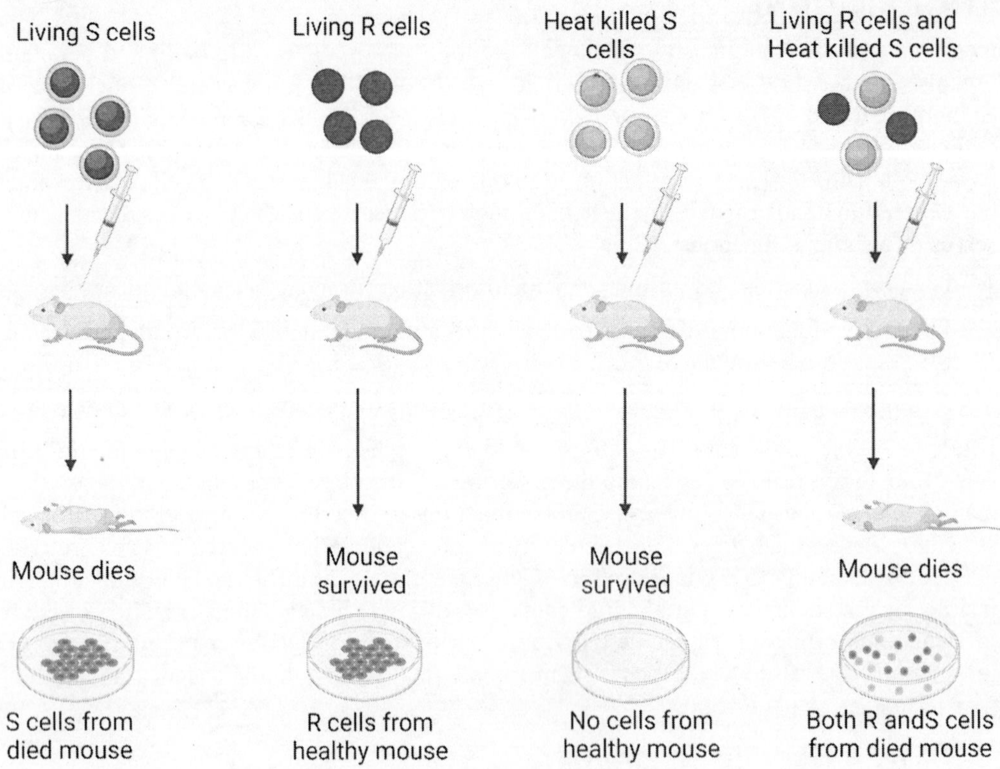

Fig. 6. 1 Griffith experiment on mice

(Images created using BioRender® software)

In this experiment (Fig. 6.1), Griffith injected these two strains of bacteria into mice. The S strain killed the mice, while the R strain did not. He further found that when heat-killed S strain was injected into mice, it did not cause pneumonia. Later, he combined the heat-killed S strain with the live R strain and injected the mixture into a mouse; the mouse developed pneumonia and died. He determined that the non-pathogenic R strain could be transformed into the pathogenic S strain if the remains of the dead virulent strain are available to the living non-virulent strain and termed it a transforming principle.

In 1944, Oswald Avery, Colin MacLeod, and Maclyn McCarty demonstrated that the transforming molecule was Deoxyribonucleic acid (DNA). A significant paradigm shift occurred due to the discovery of DNA, which led to a better understanding of the molecular basis of life.

6.1.3 Hershey-Chase Experiment

In 1952, Alfred Hershey and Martha Chase conducted an experiment identifying DNA as the genetic material of phages. Hershey shared the 1969 Nobel prize in physiology or medicine for their discovery.

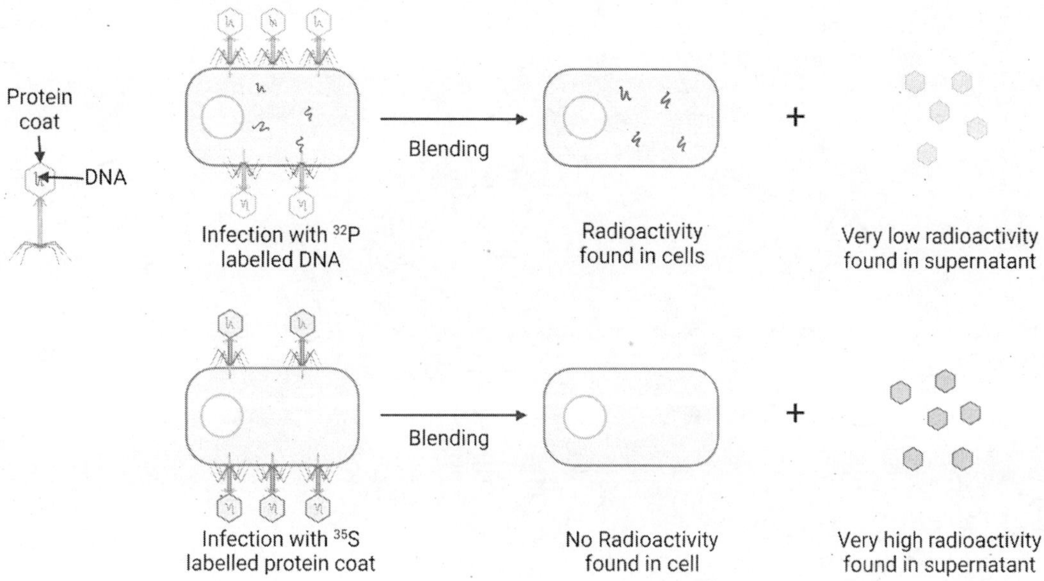

Fig. 6. 2 Hershey-Chase experiment

(Images created using BioRender® software)

In the first experiment (Fig. 6.2), they infected the bacteria with T2 phages consisting of radioactive 32P- labeled DNA. In a second experiment, bacteria were infected by T2 phages with radioactive ^{35}S-labeled protein. In both experiments, the bacteria were separated from the phage coat by blending, followed by centrifugation. In the first case, most radioactivity was found in the infected bacteria. In contrast, in the second case, most radioactivity was found in the phage coat, demonstrating that DNA is the genetic material of phages and protein does not transmit genetic information.

Although these two experiments provide the key results that DNA is the genetic material, both experiments have limitations that led to the argument that protein could still be the genetic material. In retrospect, these two experiments are essential because they alerted biologists that DNA might be the genetic material and was worth studying. This is what influenced Watson and Crick to work on DNA. Their discovery of the double-helix structure, which solved the perplexing riddle of how genes can replicate, convinced the scientific world to accept DNA as genetic material.

6.2 DOUBLE HELICAL STRUCTURE OF DNA

DNA stands for deoxyribonucleic acid, a molecule made up of four nucleotides (which consist of a nitrogenous base, a five-carbon sugar, and an ion), namely, Adenine, Guanine, Thymine and Cytosine. The nitrogenous bases, Adenine (A) and guanine (G) are double-ringed purines, and cytosine (C) and thymine (T) are single-ringed pyrimidines. The Discovery of the double helical structure of DNA was the most important breakthrough in biology during the twentieth century. The double helical structure of DNA was first described by Watson and Crick in 1953 using X-ray diffraction data of DNA fibers obtained by R. Franklin and M. Wilkins (Fig. 6.3). Watson, Crick, and Wilkins were awarded the Nobel Prize in 1962 for Medicine for discovering the molecular structure of DNA.

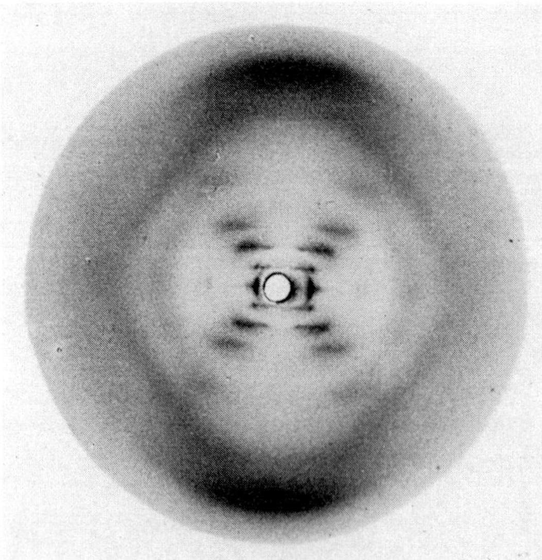

Fig. 6. 3 X-ray diffraction photo of dsDNA obtained by Rosalind Franklin

Watson and Crick proposed DNA to be made of two strands twisted around each other to form a right-handed helix, called a double helix (Fig. 6.4). The two strands are antiparallel in nature, i.e., one strand is oriented in a 5' to a 3' direction, and the other strand is oriented in a 3' to 5' direction. The strands interact by hydrogen bonds between complementary base pairs, the property of the nitrogenous bases. As a base pairing rule, adenine pairs with thymine A=T (A= U in RNA) by means of a double bond, while guanine (G) pairs through a triple bond to cytosine (C). Thus, the number of moles of A equals the number of moles of T, and the number of moles of G equals the number of moles of C. This was first found by Chargaff in the early 1950s and is called Chargaff's rule. According to Chargaff's rule, A+G=T+C or A+G/T+C=1. Where A=T and G=C. This structured base pairing results in the double-helical structure of DNA. These base pairs also serve to copy genetic information from an existing nucleic acid chain and synthesize a new chain.

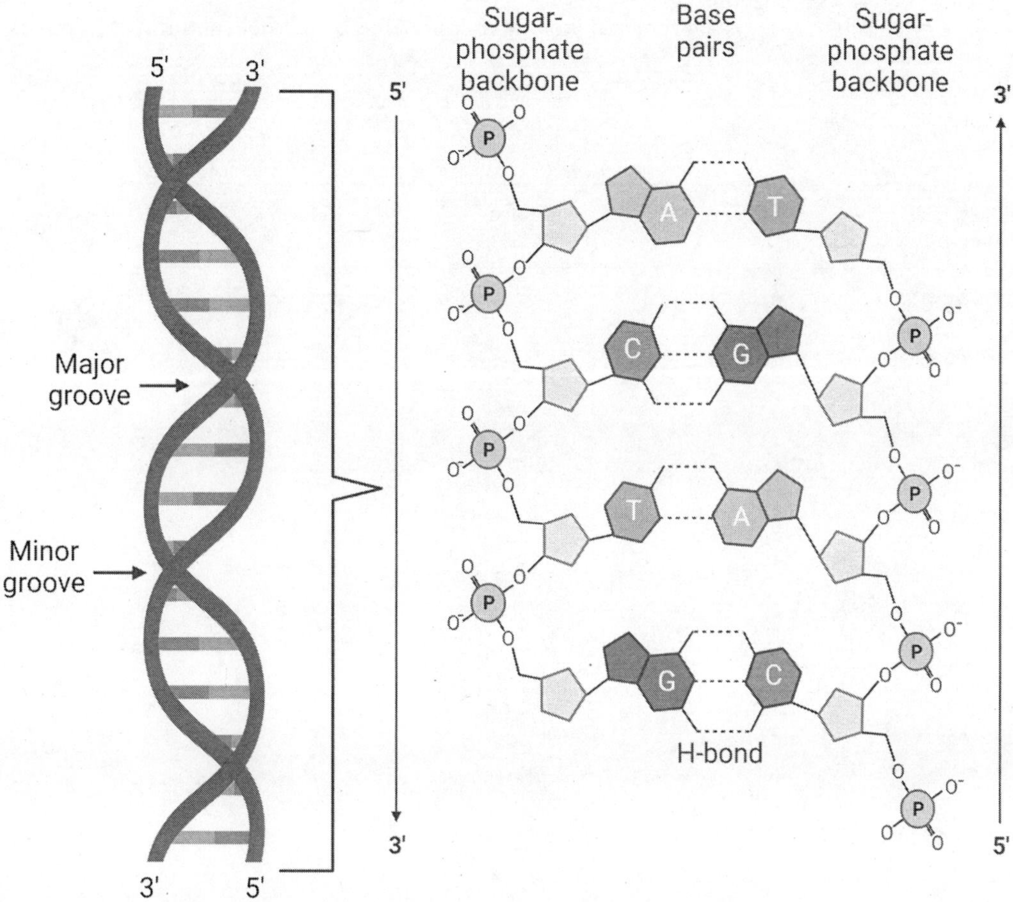

Fig. 6. 4 Double helical structure of DNA with base pairing

(Images created using BioRender® software)

Watson and Crick reported the formation of major and minor grooves because of the angle of interaction between base pairs (Fig. 6.4). The salient features of the Double-helix structure of DNA are as follows:

i. The two strands of DNA are twisted around each other to form a right-handed helix, called a double helix.

ii. The double helix consists of two intertwined strands with sugar-phosphate backbones, with the heterocyclic DNA bases projecting inwards from each of the two strands.

iii. The two strands of DNA double helix are antiparallel in nature, i.e., one strand is oriented in the 5' to 3' direction and the other in the 3' to 5' direction.

iv. H-bonding is formed between bases of opposite strands; purine bases from one strand form H-bonding with pyrimidine bases of another strand, and vice versa. Adenine forms two hydrogen bonds with thymine from opposite strands and vice-versa, while guanine forms three hydrogen bonds with cytosine. Consequently, a purine is always opposite a pyrimidine. This creates a uniform distance between the two strands of the helix.

v. The pitch of the helix is 3.4 nm, and there is roughly 10.5 bp per helical turn.

Table 6. 1 Forms of DNA

Geometry attribute	A form	B form	Z form
Helix sense	right-handed	right-handed	left-handed
Repeating unit	1 bp	1 bp	2 bp
Rotation/bp	33.6°	34.3°	60°/2
Mean bp/turn	10.7	10.5	12
Base pair tilt	20°	-6°	7°
Rise/bp along axis	2.6Å	3.4Å	3.7Å
Pitch/turn of helix	25Å	31Å	45Å
Mean propeller twist	+18°	+16°	0°
Glycosidic bond	anti	anti	anti for A, C, T, syn for G
Sugar pucker	C3'-endo	C2'-endo	A, C, T: C2'-endo, G:C3'-endo
Diameter	23Å	20Å	18Å
Major groove	Narrow and deep	Wide and deep	Flat
Minor groove	Wide and shallow	Narrow and deep	Narrow and deep

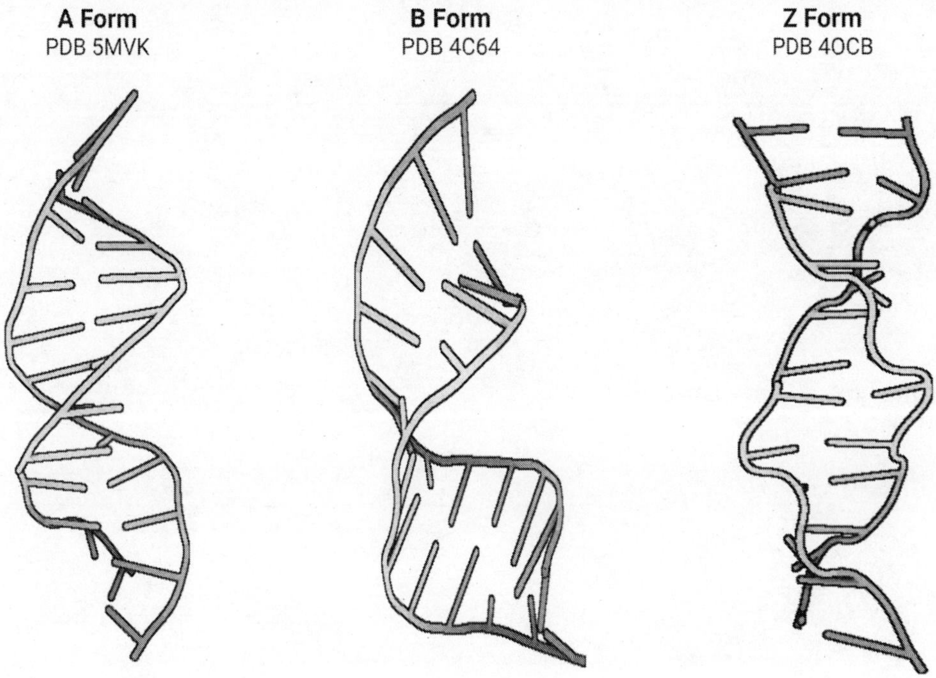

Fig. 6. 5 Structure of A, B, and Z forms of DNA

(Images created using BioRender® software)

Gene structure (Fig. 6.6) - The genetic material DNA is organized into basic physical and functional units of heredity called genes. A gene is a region of DNA encodes the synthesis of the gene product, RNA, and protein. Basically, genes in the DNA encode proteins that carry all the functions inside the cell, including metabolizing nutrients and synthesis of new cellular constituents. Prokaryotic genes lack introns (non-coding region in gene) and are organized in operons that are transcribed into RNA encoding multiple proteins (polycistronic). At the same time, eukaryotic genes are monocistronic and consist of coding exon regions interrupted by non-coding introns that are removed by splicing event post-transcription.

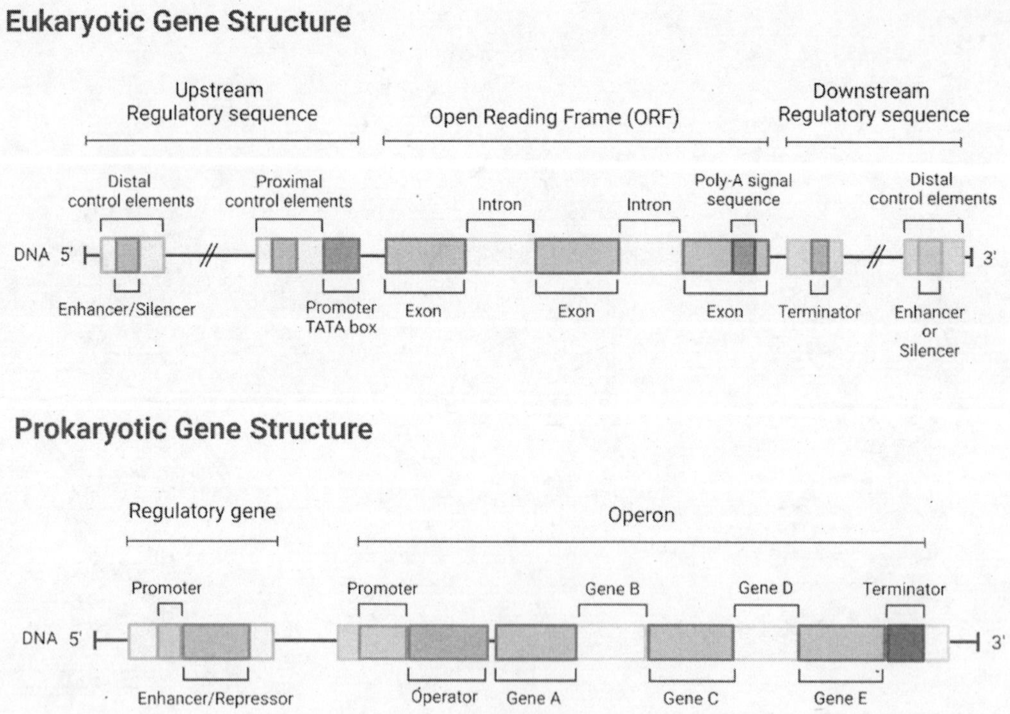

Fig. 6. 6 Gene structure

(Images created using BioRender® software)

6.3 HIERARCHY OF DNA

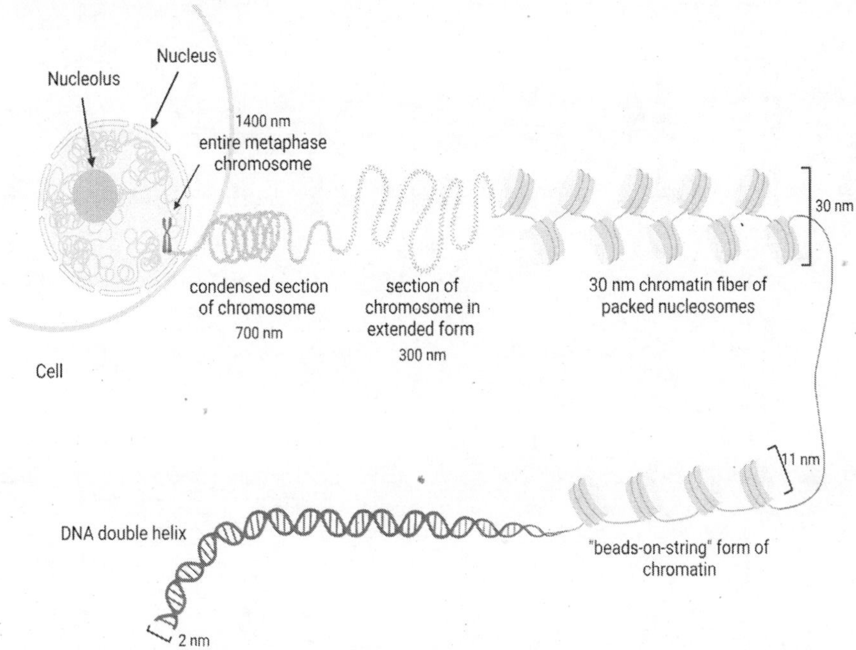

Fig. 6. 7 Hierarchies of Genome Organization

(Images created using BioRender® software)

Inside the nucleus, double-helical DNA occurs in a highly compact form (Fig. 6.7). It is packaged through a ubiquitous hierarchical process by proteins called histones. In addition to controlling the DNA's accessibility to the cellular machinery for replication, transcription, and translation, this packaging compacts the DNA to fit inside the nucleus. H2A, H2B, H3, and H4 are the four core histone proteins that play a role in packing. DNA wraps approximately 1.7 times around histone octamer, which is composed of two copies of each histone protein. This complex of histone octamer and DNA is called the nucleosome. Contiguous nucleosomes are separated by linker DNA and resemble beads-on-string. The histone protein H1 is related to linker DNA and may be present at the site where DNA enters and exits the nucleosome. There is a group of closely related additional histones collectively called linker histones. A single linker histone is attached to each nucleosome to form a chromatosome. The secondary structure of packaging involves folding this beads-on-string motif into a 30nm thick fibre called chromatin consisting of approx. 6 nucleosomes for every turn. This 30 nm chromatin fibre is majorly found in the nucleus during interphase, the period between nuclear divisions. However, when the nucleus divides, the DNA compacts forming a highly condensed metaphase chromosome. In eukaryotes, this highly condensed structure is maintained by the Structural Maintenance of Chromosome proteins, such as cohesins and condensins.

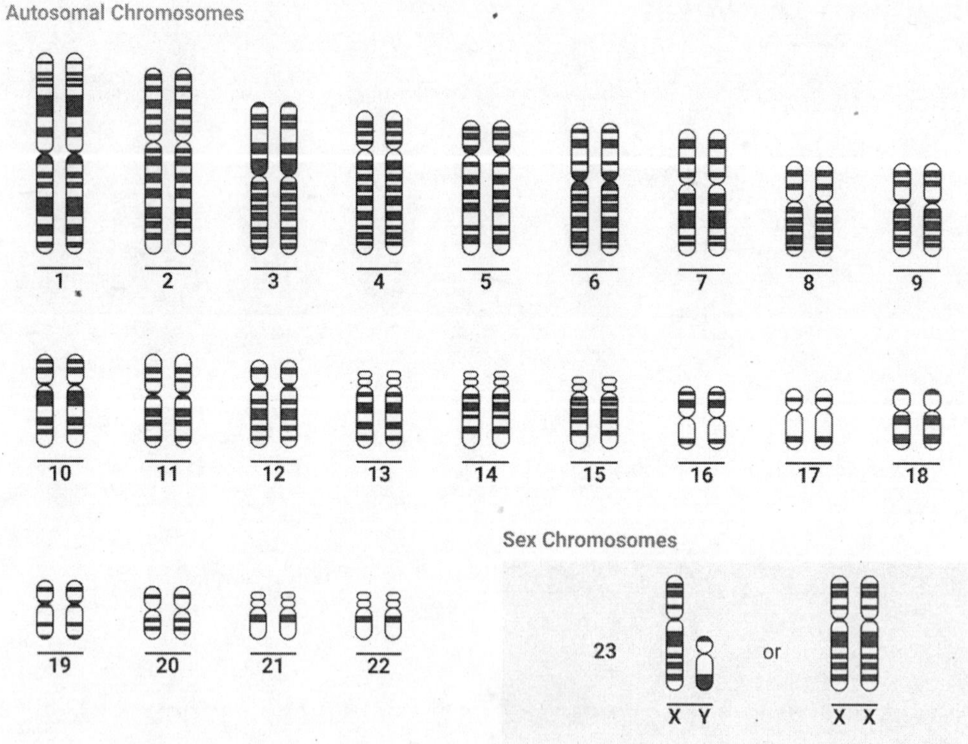

Fig. 6. 8 Human Karyotype.

(Images created using BioRender® software)

Table 6. 2 Chromosomes number in different organisms

Species	Diploid Chromosome number (2n)	Species	Diploid Chromosome number (2n)
Human	46	Potato	48
Fruit fly	8	Tobacco	24
Mouse	40	Rice	24
Rhesus	42	Corn	20
Dog	72	Pea	14
Cattle	60	Tomato	24
Chimpanzee	48	Cotton	26

6.4 MOLECULAR BASIS OF INFORMATION TRANSFER

The genetic information stored in the nucleotide sequence of DNA serves two purposes:

1. It provides the information inherited by daughter cells or offspring.
2. It is the source of information for the synthesis of all protein molecules of the cell and organism. It is called the central dogma of biological information. The Central Dogma (Fig. 6.9) explains the flow of genetic information and was proposed in 1958 by Francis Crick. It refers to the intricate base-to-base transfer of genetic information from nucleic acids to proteins via RNA. The expression of genes, meaning the synthesis of proteins, constitutes two major steps:

- Transcription - It is the synthesis of single-stranded RNA molecules from double-stranded DNA, basically a transfer of genetic information from DNA to RNA.
- Translation - It is the process of reading the mRNA sequence as the genetic code that transfers the information stored in the DNA into the amino acid sequence of proteins.

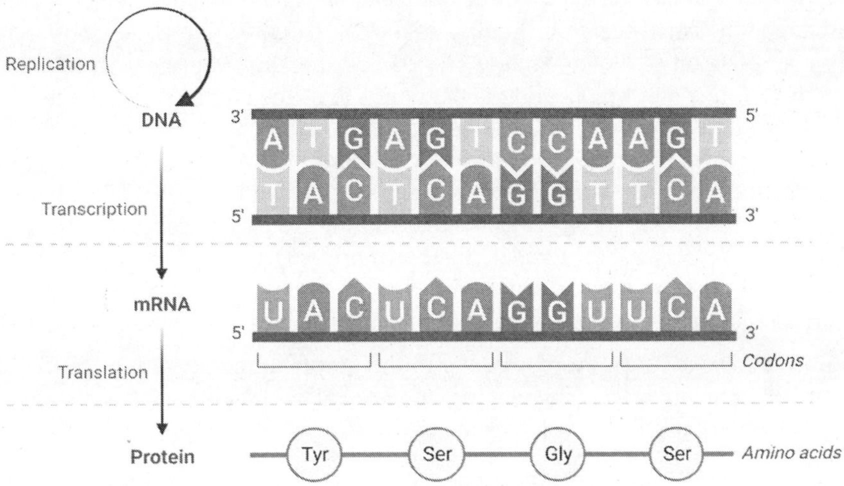

Fig. 6. 9 Central Dogma of life

(Images created using BioRender® software)

6.5 DNA REPLICATION

Cell proliferation depends on the transmission of genetic information from generation to generation. This is only possible if the genetic material, DNA, is copied accurately, resulting in two copies of the entire genome for faithful distribution into the daughter cells. This process of genome duplication is termed DNA replication. The S phase of the cell cycle is when this process takes place.

Semiconservative replication - Watson and Crick recognized that the complementary nature of the double helix plays an essential role in DNA replication. As both strands were bound together by hydrogen bonds, they could detach without breaking covalent bonds. Accordingly, the base sequence of each parental strand of the double helix would act as a template for the synthesis of the new strand. This process is called semiconservative replication, in which the parental double helix is partially conserved. Experimentally it was demonstrated by Meselson and Stahl.

Process of DNA replication - Replication starts at a particular site called the origin of DNA replication and proceeds bidirectionally or unidirectionally. A unit of DNA replication is called a replicon. Prokaryotes have single replicons and are called monorepliconic, while eukaryotes have multiple origins and are called multirepliconic. Numerous enzymes regulate this complex process of replication. The hydrogen bonds that typically hold the two strands of DNA together are broken by DNA helicase, causing the two strands of DNA to unwind or unzip, resulting in the formation of a replication fork. Primase, an enzyme associated with helicase in the primosome, creates an RNA primer extended by DNA polymerase III via complementary base pairing. By hydrogen bonding to the strand being duplicated, it inserts one additional nucleotide at a time only in the 5′ to 3′ direction. This leads to difficulty, as the top parent strand (leading strand) is replicated in the same direction as the unwinding process, but for the bottom parent strand (lagging strand), the 5′ to 3′ direction of replication is the opposite of the unwinding. This results in short pieces of DNA called Okazaki fragments, named after the biochemist who discovered them. As each Okazaki fragment is synthesized, the RNA primer of the previous strand is removed by the 5’ to 3’ exonuclease activity of DNA Polymerase I, as shown (Fig. 6.10). DNA ligase is an enzyme that joins the Okazaki fragments by forming a phosphodiester bond between them. Topoisomerase performs a crucial function during DNA replication by preventing the DNA double helix ahead of the replication fork from getting too tightly wound as the DNA unwinds. Replication terminates when the parent molecule has been completely replicated.

DNA replication is an accurate process owing to the proofreading activity of DNA polymerase. DNA proofreading involves scanning newly synthesized DNA chain for any errors and correcting them before chain extension; this is attributed to the 3’ to 5’ exonuclease activity of DNA polymerase.

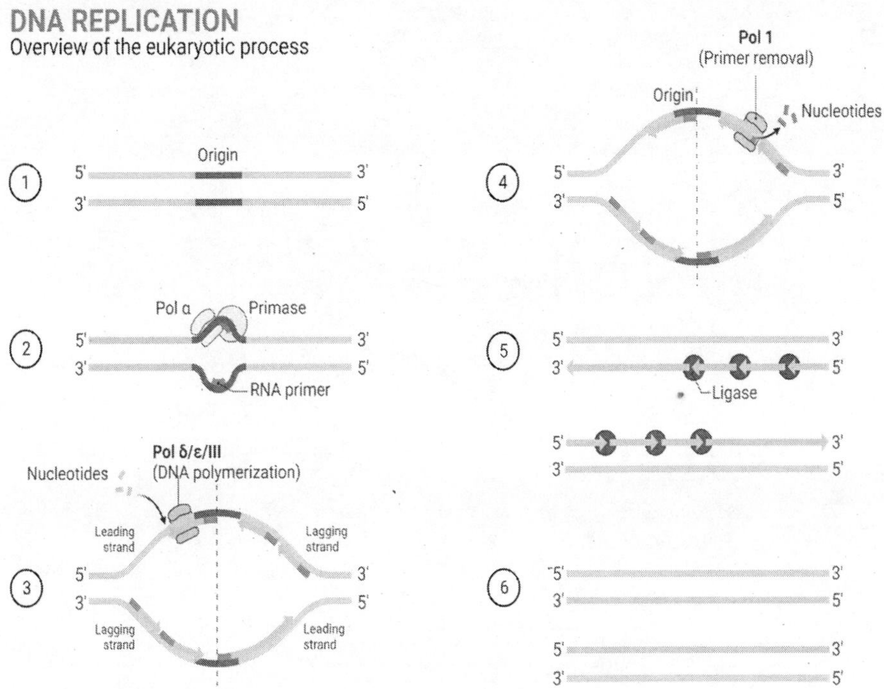

Fig. 6. 10 DNA replication

(Images created using BioRender® software)

6.6 PROTEIN SYNTHESIS

6.6.1 Transcription

Transcription is the synthesis of a single-stranded RNA chain catalyzed and scrutinized by an enzyme RNA polymerase through complementary base pairing. Transcription occurs in the 5' to 3' direction. The transcribed segment of DNA is called a transcription unit, and the newly synthesized RNA molecule is called a transcript.

Only one strand of the transcription unit is transcribed. Therefore, the transcript is identical in sequence to a strand of DNA molecule termed coding strand and complementary to the other strand that serves as a template and is termed template strand. Transcription begins when RNA polymerase attaches to the template strand and catalyzes the synthesis of RNA.

RNA Polymerase - RNA polymerases are large multi-subunit enzymes associated with different transcription factors during transcription. Bacteria possess only one RNA polymerase, whereas eukaryotes possess three distinct types, RNA pol I, which synthesizes the majority of ribosomal RNAs (rRNAs); RNA pol III, which synthesizes one small rRNA, tRNAs, as well as other small regulatory RNA molecules, while the RNA pol II synthesizes the messenger RNAs (mRNA), which serve as templates for protein synthesis.

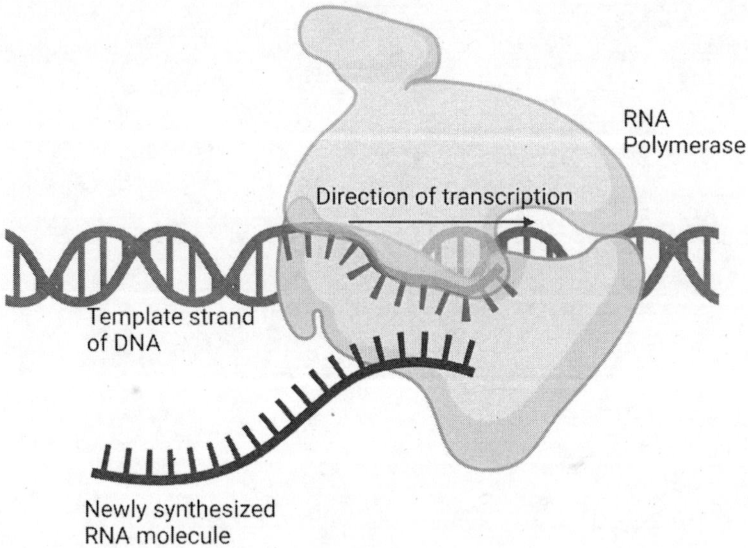

Fig. 6. 11 Process of Transcription

(Images created using BioRender® software)

Process of transcription - The transcription process (Fig. 6.11) is divided into three stages: Initiation, elongation, and termination. RNA polymerase binds to a promoter sequence on the DNA and moves along the template strand, extending its growing RNA chain in a 5' to 3' direction by stepwise addition of ribonucleoside triphosphates. Once it reaches a termination point, the newly synthesized RNA chain and polymerase are released from the DNA.

Initiation - Initiation occurs when the RNA pol attaches to the DNA sequence located 5' upstream of the gene, known as a promoter. In bacteria, promoters typically consist of three sequence components; however, in eukaryotes, there can be up to seven.

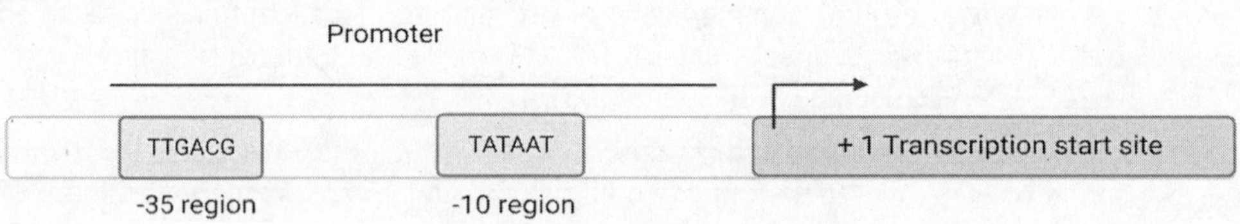

Fig. 6. 12 Prokaryotic promoter

(Images created using BioRender® software)

In most bacterial promoters (Fig. 6.12), the consensus sequence TATAAT is located 10 base pairs upstream from the transcription start site. Some strong promoters also have an upstream element, a 40-60 nucleotide A-T-rich region that increases the rate of transcription. Several genes have TTGACA sequence about 35 bases upstream of the transcription start site.

Because eukaryotes have three classes of RNA polymerase (RNA pol I, RNA pol II, and RNA pol III) that transcribe various sets of genes, eukaryotic promoters (Fig. 6.13) are more complex than their prokaryotic counterparts. Additionally, many eukaryotic genes have enhancer sequences that can influence the transcription rate. The looping of DNA brings enhancers and promoters together so that they interact with each other even though they may be thousands of nucleotides apart. This looping is caused by the interactions between the proteins associated with the enhancer and the promoter. The proteins that facilitate this looping are known as activators, whereas those that suppress it are known as repressors.

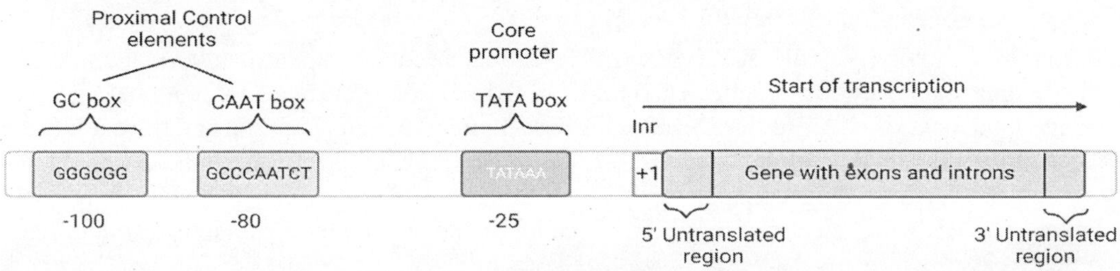

Fig. 6. 13 Eukaryotic RNA pol II promoter

(Images created using BioRender® software)

A pol II promoter consists of a consensus sequence TATTAA, often called TATA box, located about 25–35 bases upstream of the initiation site and an initiator (Inr) sequence 1 nucleotide upstream that regulates the start site and affects the transcription rate. Eukaryotic RNA polymerases require regulatory and essential cofactors known as general transcription factors. The TATA box and Inr sequences act as binding sites for general transcription factors, whereas other cis-acting areas serve as binding sites for regulatory factors. The two regulatory sequences (GC box and CAAT box) are located upstream of the core promoter and serve as a binding site for regulatory factors. Some genes also harbor downstream promoter element (DPE) located about 28–32 nucleotides downstream of the core promoter.

Promoters and enhancers are usually referred to as "strong" or "weak" according to their impact on the rate of transcription and, consequently, on gene expression. A cell might suffer negative consequences from a change in promoter strength, which frequently results in illness. For instance, some tumor-promoting viruses alter healthy cells by putting potent promoters close to genes that promote cell growth, whereas translocations in some cancer cells put genes that need to be "turned off" close to potent promoters or enhancers.

Polymerases I and III initiate the transcription of eukaryotic genes in a similar way, although the promoter sequences and transcriptional activator proteins differ.

Elongation - Upon initiation of transcription, the DNA double helix unwinds, and RNA polymerase reads the template strand, adding nucleotides to the 3′ end of the developing chain.

Termination - Terminator sequences are located near the ends of noncoding sequences. Bacteria possess two forms of these sequences. In rho-independent terminators, inverted repeat sequences are transcribed and folded back in hairpin loops, pausing RNA pol and releasing the transcript. In rho-dependent terminators, rho unwinds the DNA-RNA hybrid generated during transcription to release newly produced RNA.

Depending on the specific polymerase used, different methods are used to stop transcription in eukaryotes. A termination factor is used to halt transcription for the pol I genes in a manner reminiscent of the rho-dependent termination in bacteria. Whereas for Pol III genes, a mechanism like rho-independent prokaryotic termination is used to stop the transcription. However, pol II transcription termination is more difficult.

Pol II transcription can continue after a noncoding region for hundreds or thousands of nucleotides. A complex that appears to be associated with the polymerase then cleaves the RNA strand. The 3′ ends of mature pol II mRNAs are polyadenylated, resulting in a poly(A) tail; this process is coordinated with cleavage and termination.

In addition to the poly(A) tail, eukaryotic mRNAs feature a 5' cap structure made up of 7-methylguanosine residues connected by a 5'-5' triphosphate bridge. During transcription, 7-methylguanosine is added to the 5' end of the nascent mRNA. Transcription and translation occur in distinct cellular compartments in eukaryotes: transcription in the nucleus and translation in the cytoplasm. Before being transported to the cytoplasm for translation, the main transcript of eukaryotic RNA polymerase is modified (5' capping, polyadenylation, and intron splicing).

6.6.2 Translation

During translation, the information stored in messenger RNA (mRNA) regulates the incorporation of amino acids during protein synthesis. Translation occurs on ribosomes in the cell cytoplasm, where mRNA is read and translated into amino acid chains.

Three forms of RNA molecules serve distinct but complementary roles in protein synthesis:

Messenger RNA (mRNA) - There are two sorts of regions in mRNAs. The coding region begins with a start codon AUG and ends with a termination codon. Except for the coding region, the mRNA molecule also consists of extra regions at both ends. The 5' and 3' untranslated regions are known as the leader and trailer, respectively. Intercistronic sections can also be seen in polycistronic mRNAs.

Transfer RNA (tRNA) - tRNA is essential for decoding the information in mRNA. tRNA (70–80 nucleotides) has a sequence that enables hairpins to form. This gives a secondary structure to the molecule. The anticodon forms one end of the molecule when the tRNA folds, while the other end will be joined to a certain amino acid. An anticodon is a group of three nucleotides located in the center of the tRNA. Each type of amino acid has a unique tRNA that binds it and transports it to the expanding polypeptide chain if the subsequent mRNA code specifies it.

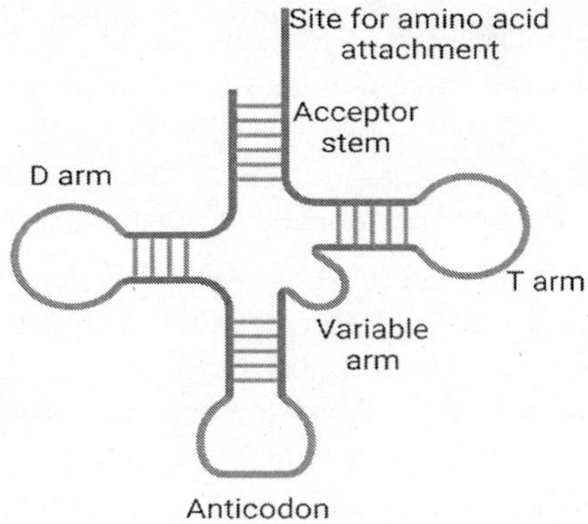

Fig. 6. 14 Structure of tRNA

(Images created using BioRender® software)

Ribosomal RNA (rRNA) - rRNA, along with a set of proteins, forms ribosomes. In order to catalyze the incorporation of amino acids into protein chains, these intricate structures physically move along mRNA in the 5' to 3' direction. Additionally, they bind tRNAs and other support molecules needed for protein synthesis.

Ribosomes – This are the molecular workbench of translation. They are ribonucleoproteins containing rRNA and ribosomal proteins (r-proteins) and are made up of two subunits. In prokaryotes, the large subunit is the 50S and the small subunit is 30S, while in eukaryotes, the large subunit is 60S, and the small subunit is 40S.

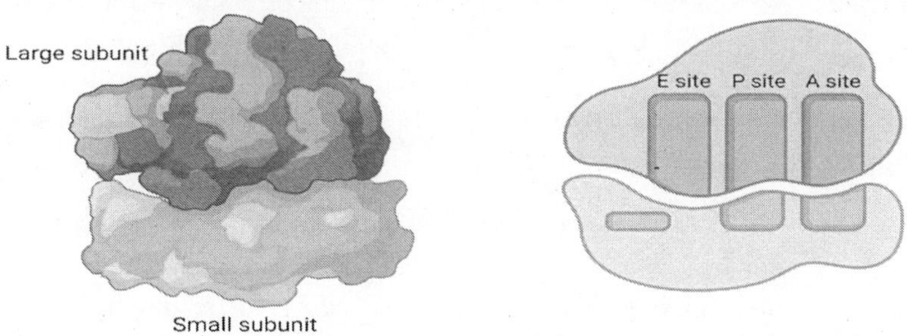

Fig. 6. 15 Structure of the ribosome with its binding sites

(Images created using BioRender® software)

The ribosome has three binding sites for tRNA (Fig. 6.15):

P-site: Also known as the peptidyl-tRNA binding site, binds the tRNA molecule to the polypeptide chain's expanding end.

A-site: Also known as the aminoacyl-tRNA binding site, it is responsible for holding the incoming tRNA molecule that is charged with an amino acid.

E-site: Deacylated tRNA devoid of amino acids exits via the E site, also known as the exit site.

The ribosome reads the continuous chain of three-nucleotide groups and constructs the protein in accordance with the instructions found in the mRNA sequence. Proteins typically have a length of several hundred amino acids or more. The ribosome will leave the mRNA once a complete protein has been constructed and search for the 5' end of another mRNA to start the translation. One copy of a protein is produced each time one ribosome interprets an mRNA.

Activation of amino acid and attachment to tRNA - The first step in translation is activating an amino acid by ATP and its attachment to its corresponding tRNA through a process known as tRNA charging or aminoacylation. Aminoacyl tRNA synthetases are a group of enzymes responsible for the attachment of amino acids to their corresponding tRNA. Almost all species possess 20 aminoacyl-tRNA synthetases, one for each amino acid.

Genetic Code

General features of genetic code:

1. The genetic code is a triplet code called a codon, meaning a triplet of nucleotides codes for a specific amino acid.
2. Each triplet specifies a specific amino acid; thus, the coding is unambiguous.
3. The code consists of start and stop signals, which are required to initiate and terminate translation. Three codons out of 64 do not code for any amino acids and are known as stop codons (UAA, UAG, and UGA), while the code AUG acts as a start codon and codes for methionine.
4. Once the translation of mRNA commences, codons are read consecutively with no gaps in between. Thus, the code is said to be comma less.
5. The code is degenerate. A particular amino acid can be specified by more than one triplet codon. The difference between them lies at the third position, while the first two bases of the codon determine specificity. This is true for 18 of the 20 amino acids. The different codons for a particular amino acid are synonymous. *The Wobble hypothesis explains the degeneracy of the code*; this occurs because tRNAs for some amino acids include the nucleotide inosinate, which forms weaker interactions with the third base of the codon. Different codons for the amino acid are shown in Fig. 6.16. The wobble in the codon not only contributes to specificity but also permits rapid dissociation of the tRNA from its codon, thus balancing the accuracy and speed of protein synthesis.
6. The code is non-overlapping.
7. The genetic code is nearly universal, meaning the code remains the same for almost all organisms with few exceptions. These expectations have been found in the genome of protozoans, mycoplasma, and mitochondria. For example, from bacteria to humans, UGA is a stop codon, but in mitochondrial DNA and *Mycoplasma capricolum,* it codes for tryptophan. Another variation occurs in ciliated protist, where both the termination codons, UAG and UAA, codes for glutamine.

Second base in codon

First base in codon	U	C	A	G	Last base in codon
U	UUU Phe	UCU Ser	UAU Tyr	UGU Cys	U
U	UUC Phe	UCC Ser	UAC Tyr	UGC Cys	C
U	UUA Leu	UCA Ser	UAA STOP	UGA STOP	A
U	UUG Leu	UCG Ser	UAG STOP	UGG Trp	G
C	CUU Leu	CCU Pro	CAU His	CGU Arg	U
C	CUC Leu	CCC Pro	CAC His	CGC Arg	C
C	CUA Leu	CCA Pro	CAA Gln	CGA Arg	A
C	CUG Leu	CCG Pro	CAG Gln	CGG Arg	G
A	AUU Ile	ACU Thr	AAU Asn	AGU Ser	U
A	AUC Ile	ACC Thr	AAC Asn	AGC Ser	C
A	AUA Ile	ACA Thr	AAA Lys	AGA Arg	A
A	AUG Met (start)	ACG Thr	AAG Lys	AGG Arg	G
G	GUU Val	GCU Ala	GAU Asp	GGU Gly	U
G	GUC Val	GCC Ala	GAC Asp	GGC Gly	C
G	GUA Val	GCA Ala	GAA Glu	GGA Gly	A
G	GUG Val	GCG Ala	GAG Glu	GGG Gly	G

Fig. 6. 16 The genetic code dictionary

(Images created using BioRender® software)

Codon bias: The term "codon bias" describes the fact that not every codon is utilized with equal frequency in the genes of a specific organism. Among the four valine codons, human genes utilize GTG four times more often than GTA. The biological cause of codon bias is unknown, but it has been established that codon bias directly affects protein expression and regulation and influences protein folding. Although all organisms have a codon bias, it varies between species.

Process of translation - The process involves three main steps:

Initiation - In prokaryotes, the small (30S) and large (50S) subunit associates at the ribosome binding site, a short sequence preceding the coding region. The ribosome binding site is positioned with 10 base pairs (bp) upstream of the initiation codon. It has the sequence 5'-AGGAGGU-3' (also known as the Shine-Dalgarno sequence) complementary to the 3' end of 16S rRNA. In eukaryotes, the small subunit identifies the methylation cap at the 5' end of mRNA. It scans the sequence until it reaches the initiation codon, where it is joined by the large subunit. AUG is the initiation codon in most cases, but in E. coli, GUG and rarely UUG also serve as the initiation codon. In eukaryotes, the initiation codon AUG is surrounded by a few nucleotides called the Kozak sequence 5'-ACCAUGG-3'. Initiation factors play a role in translation initiation. In bacteria, there are 3 initiation factors, IF-1, IF-2, and IF-3, whereas eukaryotes have 12 factors and also require a poly(A) tail for the formation of the initiation complex at the 5' end. The initiation codon codes for methionine, so protein synthesis begins with the amino acid methionine. However, in *E. coli* and other prokaryotes, it is N-formylmethionine.

Elongation - During elongation (Fig. 6.17), peptide bonds are synthesized between amino acids of the expanding polypeptide chain. One amino acid is added to the developing peptide chain at a time on the ribosome during this cyclic process, aided by elongation factors. Ribosomes elongate polypeptide chain in three stages:

Decoding: At the A site, the ribosome attaches to an aminoacyl-tRNA with an anticodon complementary to the mRNA codon;

Transpeptidation: By forming a peptide bond, the peptidyl group on the P site of tRNA is transferred to the aminoacyl group at the A site;

Translocation: Newly produced peptidyl tRNA at the A site is transported to the P site, while uncharged tRNA at the P site enters the E or exit site of the ribosome. The ribosome advances exactly three nucleotides in the 5' to 3' direction along the mRNA molecule. This process of elongation in bacteria is aided by three elongation factors.

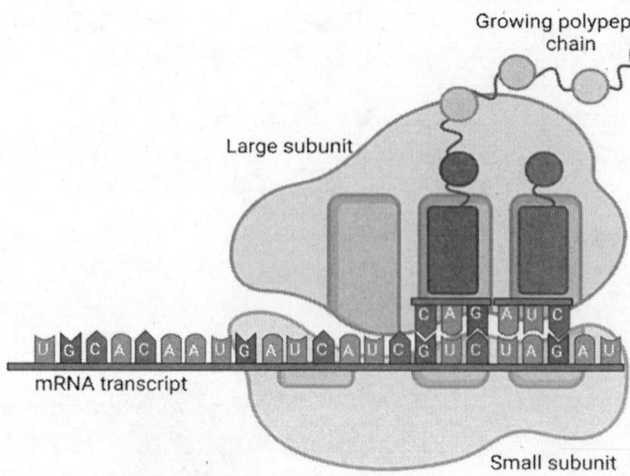

Fig. 6. 17 Protein synthesis on ribosomes

(Images created using BioRender® software)

Termination - Termination involves the release of the newly synthesized polypeptide chain and the dissociation of the ribosome from the messenger RNA when one of the three termination codons (UAG, UAA, and UGA) is reached. This is brought about by release factors, RF I, RF II, and RF III, which enter the A site of ribosome instead of tRNA.

6.7 GENE EXPRESSION AND REGULATION

A Gene expression is aided by the process of transcription and translation, during which DNA-encoded information is decoded into proteins. It is a tightly regulated process that helps the cells to respond to the changing environment. Cells control protein synthesis through various proteins involved in regulating gene expression, called regulatory proteins.

These regulatory proteins either act as an activator or repressor of gene expression. Basically, an activator bound to its target site (operator) increases the rate of transcription while the binding of the repressor represses the expression of that gene. A classic example of gene regulation is the *Lac* operon model.

6.7.1 Lac Operon Model

Lac operon was the first operon to be discovered, consisting of three structural genes, an operator, and a regulatory gene. The structural genes of *Lac* operon (Fig. 6.18) code for the enzymes involved in lactose metabolism.

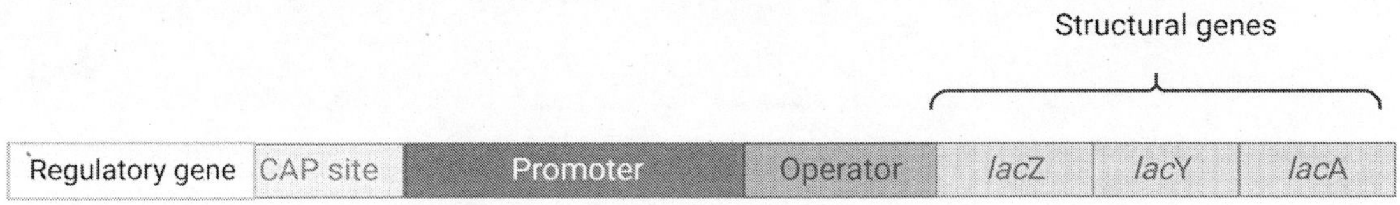

Fig. 6. 18 Structure of lac operon

(Images created using BioRender® software)

Regulatory gene (I): The regulatory gene codes for the regulatory protein, *lac* repressor, which controls the transcription of *lac* operon by binding to the operator.

***Lac* operator** (O): The palindromic sequence of 26 bp between the promoter and the structural genes interacts with the lac repressor or a regulatory protein that controls the transcription of the structural genes.

Lac structural genes: *Lac* operon consist of three structural genes.

***lac*Z**: codes for an enzyme β-galactosidase that breaks down *lac*tose into glucose and galactose. Additionally, it catalyzes the conversion of lactose into allolactose.

***lac*Y**: codes for an enzyme β-galactosidase permease that helps in lactose transport through the cell membrane.

***lac*A**: codes for β-galactosidase transacetylase that catalyzes the transfer of the acetyl group from acetyl-CoA to β-galactosidases. Though it is not involved in lactose metabolism, it is involved in the detoxification of compounds.

All the genes of the *lac* operon are transcribed into a multigenic mRNA molecule. The *lac* operon is an inducible operon expressed in the presence of an inducer. Allolactose, formed from lactose by the action of β-galactosidase, acts as an inducer of the *lac* operon. The relief of repression by an inducer is termed induction.

Negative regulation of lac operon

1. In the absence of an inducer, the *lac* repressor attaches to the operator preventing RNA polymerase from initiating transcription at the promoter. The *lac* repressor molecule is a negative regulator of the *lac* operon and acts as a roadblock on the DNA (Fig. 6.19).

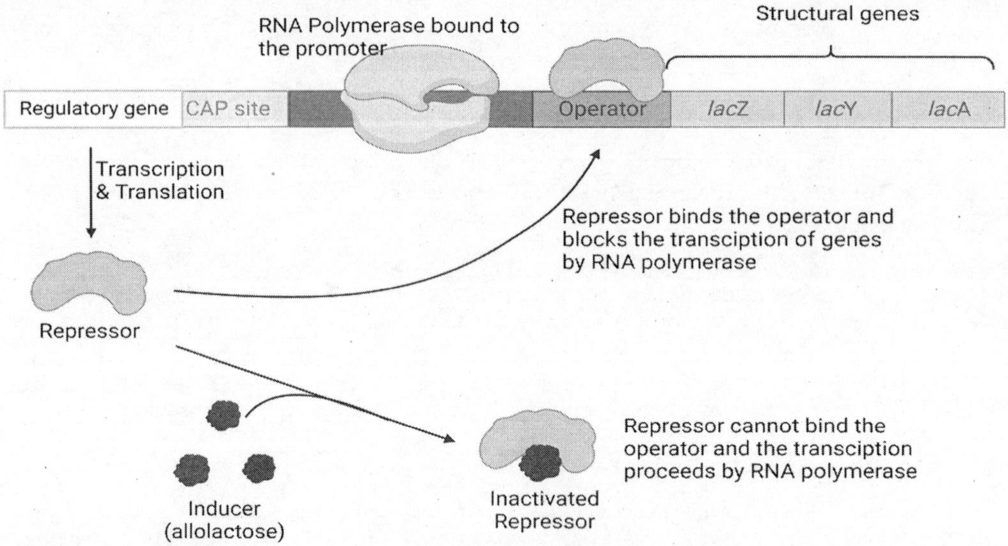

Fig. 6. 19 Regulation of Lac operon

(Images created using BioRender® software)

2. The operator binding site of the lac repressor is rendered inactive when the inducer binds to its allosteric site. This inactivation induces the transcription of structural genes of the lac operon. All three structural genes are transcribed as a single polycistronic mRNA that is then processed and translated into the enzymes β-galactosidase, permease, and transacetylase.

Sometimes an analog of lactose is capable of inducing *lac* operon without serving as a substrate for β-galactosidase and is called a gratuitous inducer; an example is isopropylthio galactoside (IPTG). However, constitutive expression of *lac* operon occurs in cases where the lacI gene has been mutated, leading to a product incapable of binding to the operator or where mutation occurs in the *lac* operator region such that the repressor molecule cannot bind and inhibit the transcription of the *lac* operon.

Positive regulation of lac operon: Catabolite repression

Bacteria have been known to show preferences with regard to carbon sources. In *E. coli,* it has been observed that glucose is preferred over lactose. When both glucose and lactose are present in the media, the organism metabolizes the glucose first and represses the lactose metabolism, leading to diauxic growth. This is termed as glucose effect and was recognized in bacteria by Monod. Due to the suppressive effect of glucose on the production of enzymes required for the metabolism of other sugars, the term catabolite repression was coined to refer to the glucose effect.

Catabolite repression is caused by cyclic AMP (cAMP) and cyclic AMP receptor protein (CRP), also known as Catabolite Activator Protein (CAP). Cyclic AMP is synthesized by an enzyme adenylate cyclase, whose concentration depends on glucose concentration. When the glucose concentration in the cell is high, the cAMP concentration is low. Conversely, when glucose concentration is low, the cAMP concentration is high. The cAMP and CRP bind together to form a complex that binds to a site adjacent to the promoter of the *lac* operon. The binding of the cAMP-CRP complex introduces a bend in the DNA, allowing CRP to contact RNA polymerase and start transcription. Thus, the cAMP-CRP complex is a positive regulator of the *lac* operon.

Table 6. 3 Catabolite repression

Glucose concentration	Lactose concentration	cAMP concentration	*Lac* promoter and operator region	Transcription of the *lac* operon
Low	Low	High	Bound by cAMP-CRP complex and *lac* repressor	Low
High	Low	Low	Bound by the *lac* repressor	Low
High	High	Low	Both *lac* repressor and cAMP-CRP complex are not bound	Low
Low	High	High	Bound by the cAMP-CRP complex	High

6.8 DNA RECOMBINATION

Genetic recombination is the exchange of genes between chromosomes. There are three classes of genetic recombination: Homologous recombination, site-specific recombination, and DNA transposition. Recombinational rearrangement of DNA sequences is a tightly coordinated process with DNA replication and repair.

Homologous recombination - In homologous recombination, genetic exchange occurs between two DNA molecules that share identical (homologous) sequences. Homologous recombination is primarily a DNA repair process in bacteria, whereas, in eukaryotes, it also plays a role in replication and cell division. In higher organisms, this process occurs during meiosis between sister chromatids; this exchange of genetic material is also known as crossing over. The model for homologous recombination was first described by Robin Holliday in 1964. Two aligned homologous DNA duplexes are nicked, and the nicked strands cross across to pair with the nearly complementary strands of the homologous duplex, yielding a Holliday junction (Fig. 6.20). The crossover point can migrate in either direction, often by thousands of nucleotides, in a process known as branch migration, in which the four strands of DNA duplex switch their base-pairing partners. The Holliday junction can be resolved in two ways. If cleaved vertically, it results in an exchange of DNA, but recombination does not occur if cleaved horizontally.

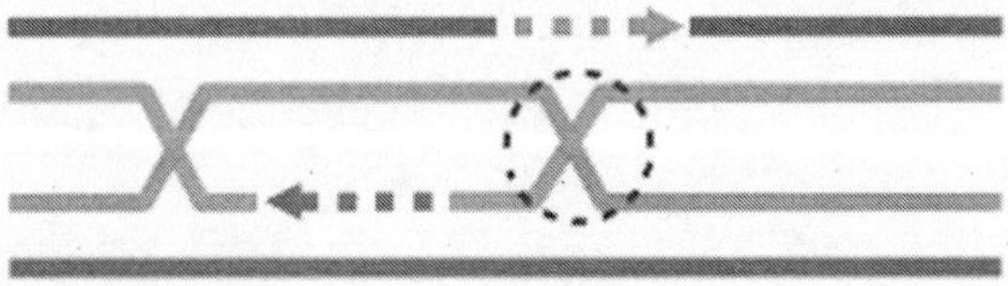

Fig. 6. 20 Holliday junction formation, Double crossover

(Created with BioRender.com)

In the double-strand break repair model of recombination, the 5' ends of the DNA strands are degraded by exonucleases forming 3' single-strand overhangs. The 3' strand overhang then pairs with the complementary strand on the intact homolog while the other strand of the intact duplex is displaced. The invading 3' end acts as a primer for DNA replication by DNA polymerase and, with branch migration, eventually generates a DNA molecule with two crossovers called Holliday intermediates.

Site-specific recombination - This recombination occurs only at particular DNA sequences. Site-specific recombination involves an enzyme called recombinase and a short DNA sequence where the recombinase acts (recombination site). Recombinases cleave the DNA at the recombination site and ligate the strands to new partners. This process can also involve a Holliday intermediate. This type of recombination is found in all cells, and its main functions include DNA integration and regulation of gene expression.

Transposition - This type of recombination involves a short segment of DNA called a transposon, capable of moving from one location in the chromosome to another. It also consists of a protein transposase coded by the transposon that promotes the process. It was first identified by Barbara McClintock while working on maize and was called jumping genes. Homology is not required in transposition, and the site of insertion is more or less random.

In bacteria, transposition occurs via two routes: Direct or simple transposition, wherein the transposon is excised and moves to a new location, and replicative transposition, wherein the entire transposon is replicated and then transported to a new site leaving behind a copy at the donor site. Eukaryotes also have transposons similar in structure to prokaryotic transposons. An example is a B-lymphocyte differentiation in vertebrates, wherein a controlled recombination reaction related to transposition unites immunoglobulin gene fragments to produce immunoglobulin genes.

UNIT SUMMARY

Nucleic acids are made up of long polymers of nucleotides. Although DNA and RNA are genetic building blocks, DNA is chemically and structurally more stable. Inside the nucleus, the double-helical structure of DNA occurs in a highly compact form called a chromosome. The hydrogen bonding between the nucleotides from opposing strands is the distinguishing feature of DNA's double helical structure. According to the rule, adenine pairs with thymine through two H-bonds, while guanine pairs with cytosine through three H-bonds. As a result, one strand becomes complementary to the other. The complementary H-bonding controls the semi-conservative DNA replication process, which occurs in the nucleus through the involvement of enzymes, DNA polymerase, helicase, and ligase. Any damage to the DNA is repaired by various mechanisms inside the cell. A segment of DNA that codes for RNA is referred to as a gene. Additionally, one of the DNA strands serves as a template to guide the production of complementary RNA during transcription. The transcribed mRNA in bacteria can be translated directly since it is functional. Eukaryotic mRNA undergoes processing such as intron splicing to create functional RNA. The triplet genetic code found in the messenger RNA is read by an adaptor molecule called tRNA using the complementarity principle. Every amino acid has a distinct tRNA. The tRNA associates with a particular amino acid at one end and forms H-bonds with anticodons on the mRNA to match with the codes there. Ribosomes attach to mRNA and serve as a platform for the synthesis of polypeptide chains of amino acids (also known as protein synthesis). An example of an RNA enzyme is one of the rRNAs, which serves as a catalyst for the creation of peptide bonds (ribozyme). The fact that translation is a mechanism that developed around RNA suggests that RNA was the building block of life. Since transcription and translation require a lot of energy, they must be strictly controlled. The first stage in controlling gene expression is the regulation of transcription. In bacteria, several genes are organized and controlled in groups called operons. The lac operon, which codes for the genes involved in lactose metabolism, is the classic example of an inducible operon found in bacteria, as the quantity of lactose in the media used to grow the bacteria controls the operon. As a result, this regulation can also be thought of as the regulation of enzyme synthesis by its substrate.

EXERCISES

Multiple Choice Questions

1) Who first showed that DNA was the transforming principle?
   A. Griffith
   B. Franklin
   C. Chargaff
   D. Hershey

2) Which scientist used viruses and radioactive isotopes to prove that DNA, not protein, was the molecule of heredity?
   A. Griffith and Avery
   B. Watson and Crick
   C. Hershey and Chase
   D. Chargaff and Franklin

3) The double helical structure of DNA was explained by
   A. Watson and Crick
   B. Rosalind Franklin
   C. Freidrich Miescher
   D. Erwin Chargaff

4) What are the four nitrogenous bases in DNA
   A. Adenine, Guanine, Glucose, and Thymine
   B. Adenine, Uracil, Cytosine, and Guanine
   C. Guanine, Glycerol, Cytosine, and Thymine
   D. Adenine, Guanine, Thymine, and Cytosine

5) Nucleotides are made up of sugar, phosphate and ____
   A. Fatty acid
   B. Nucleic acid
   C. Nitrogenous base
   D. Cholesterol

6) What is the 30 nm Bead on string structure called as
   A. Chromosome
   B. Nucleosome
   C. Chromatin
   D. DNA

7) The proteins that play a role in the packaging of DNA are
   A. Histones
   B. Centromeres
   C. Nucleosome
   D. Chromatids

8) The most compact form of DNA is called as
   A. DNA double helix
   B. Chromatin
   C. Chromosome
   D. Nucleosome

9) Where is the genetic material stored
   A. Cytosol
   B. Vacuole
   C. Cell membrane
   D. Nucleus

10) The process of making a copy of DNA is called
   A. Transcription
   B. Replication
   C. Translation
   D. Expression

11) During replication, which enzyme adds complementary bases
   A. Helicase

B. DNA Polymerase
C. Ligase
D. None of the above

12) DNA replication occurs in which direction
A. 3' to 5'
B. 5' to 3'
C. Both a and b
D. None of the above

13) An enzyme that unzips the DNA and prepares it for replication
A. Helicase
B. Topoisomerase
C. DNA Polymerase
D. Ligase

14) Okazaki fragments are formed on the
A. Lagging strand
B. 5'-end
C. Leading strand
D. Klenow fragment

15) Physical evidence of recombination was given by
A. McClintock and Creighton
B. Watson and Crick
C. Hershey and Chase
D. Freidrich Miescher

16) The process by which the genetic code is DNA is copied into RNA is called
A. Transcription
B. Replication
C. Translation
D. Transformation

17) Transcription is carried out by an enzyme
A. RNA polymerase
B. DNA polymerase
C. Helicase
D. Ligase

18) Order of genetic information transfer
A. Protein to DNA to RNA
B. RNA to DNA to protein
C. DNA to protein to RNA
D. DNA to RNA to protein

19) Translation occurs in _____ of eukaryotes
   A. Nucleus
   B. Cytoplasm
   C. Mitochondria
   D. Peroxisome

20) Complex of protein and RNA that plays a role in translation
   A. Nucleosome
   B. Ribosome
   C. Peroxisome
   D. Polymerase

21) Triplets of mRNA read by ribosomes are called
   A. Codons
   B. Anticodons
   C. rRNA
   D. Gene

22) Which of the following is a start codon
   A. UAA
   B. UGA
   C. AUG
   D. AAG

23) Which of the following is a stop codon
   A. UAA
   B. AUG
   C. GUG
   D. GGG

24) Type of RNA that carries amino acids
   A. mRNA
   B. tRNA
   C. rRNA
   D. snRNA

25) One way to regulate the expression of genes is to control the formation of mRNA. This is called ____ control
   A. Translational
   B. Post-transcriptional
   C. Post-translational
   D. Transcriptional

26) Protein that binds to the operator and blocks RNA polymerase

A. Inducer
B. Activator
C. Repressor
D. Enhancer

27) Inducer of *Lac* operon?
A. Allolactose
B. IPTG
C. Glucose
D. Both a and b

Answers: 1)A; 2)C; 3)A; 4)D; 5)C; 6)C; 7)A; 8)C; 9)D; 10)B; 11)B; 12)B; 13)A; 14)A; 15)A; 16)A; 17)A; 18)D; 19)B; 20)B; 21)A; 22)C; 23)A; 24)B; 25)D; 26)C; 27)D

Short Answer Type Questions

1) State the process of information transfer.
2) Difference between DNA and RNA.
3) Hierarchy of DNA organization.
4) Define genes.
5) What do you mean by DNA replication?
6) State different types of DNA repair
7) Define transcription.
8) Define translation.
9) What is mRNA?
10) What is tRNA, and how are they activated?
11) What are ribosomes?
12) What is Chargaff's rule?
13) What do you mean by codon bias?
14) What is an inducer?
15) What do you mean by repressor?
16) State the genes encoded by the *Lac* operon.

Long Answer Type Questions

1) Explain Griffith's experiment in brief.
2) Explain the double helical structure of DNA.
3) Explain the process of DNA replication.

4) Explain different types of DNA repair.
5) Describe the process of transcription.
6) State the salient features of the genetic code.
7) Describe the process of translation.
8) Explain the *Lac* operon model in brief.

NUMERICAL PROBLEMS

1) Transcribe the given sequence of DNA into mRNA: 5'- ATGCTACGTATAACGCATT -3'
2) If the sequence of one strand of DNA is written as follows: 5'-ATTGCTACGTTATCTGAACTTGAT -3'. Write the sequence of the complementary strand in the 5'-3' direction.

PRACTICAL

Isolation of genomic DNA

Requirements: Water, 91% Isopropyl alcohol, dishwashing liquid, salt, plastic cup

Experimental procedure:

1. Prepare a saline solution by adding 1 tbsp salt to the water. Stir until the salt is dissolved.
2. Vigorously gargle salt water and spilt it into a plastic cup. This will cause your cells to come into the solution.
3. Add dishwashing solution to the salt solution containing cells. The detergent will break open the cells.
4. Now, slowly add isopropyl alcohol to the solution such that a layer of isopropyl alcohol is formed at the top.
5. As DNA is insoluble in isopropyl alcohol, it will precipitate and become visible as a white thread in the solution after a few minutes.

KNOW MORE

Epigenetics involves genetic control without any change in DNA sequence. Epigenetic information is not encoded in the DNA. Still, it is passed from one generation to the next, mostly in the form of covalent modification of histones and/or the placement of histone variants in chromosomes. The chromatin regions with highly active gene expression (euchromatin) consist of histone variants H3.3 and H2AZ in place of histones H3 and H2A, respectively.

Expansion of genetic code - There are two extra amino acids, selenocysteine, and pyrrolysine, that are specified by the genetic code. In total, there are 22 rather than 20 amino acids specified by the known genetic code. The two extra amino acids are found in only a very few proteins and offer insights into the complexity of code evolution.

REFERENCES AND SUGGESTED READINGS

- Russell, P.J. and Gordey, K., 2002. *Genetics* (No. QH430 R87). San Francisco: Benjamin Cummings.
- Voet, D. and Voet, J.G., 2010. *Biochemistry*. John Wiley & Sons.
- David, L., Nelson, D.L., Cox, M.M., Stiedemann, L., McGlynn Jr, M.E. and Fay, M.R., 2000. Lehninger principles of biochemistry.

Dynamic QR Code for Further Reading

7 Macromolecular Analysis

UNIT SPECIFICS

Through this unit we have discussed the following aspects:

- *Reductionism approach to learning complex bio-systems*
- *Understanding how various analytical methods exploit properties of protein to estimate the concentration of proteins*
- *Describe the multiple forces that stabilize protein structures and the hierarchy in protein structures*
- *Introduction of proteins as receptors, transporters, structural elements, and biocatalysts*

The ***practical applications of the topics are discussed*** *for generating further curiosity and creativity as well as improving problem solving capacity.*

Besides giving a large number ***of multiple choice questions as well as questions of short and long answer types*** *marked in two categories following lower and higher order of* ***Bloom's taxonomy****, assignments through a number of numerical problems, a list of references and suggested readings are given in the unit so that one can go through them for practice. It is important to note that for getting more information on various topics of interest some* ***QR codes*** *have been provided in different sections which can be scanned for relevant supportive knowledge.*

After the related practical, based on the content, there is a ***"Know More"*** *section. This section has been carefully designed so that the supplementary information provided in this part becomes beneficial for the users of the book. This section mainly highlights the initial activity, examples of some interesting facts, analogy, history of the development of the subject focusing the salient observations and finding, timelines starting from the development of the concerned topics up to the recent time, applications of the subject matter for our day-to-day real life or/and industrial applications on variety of aspects, case study related to environmental, sustainability, social and ethical issues whichever applicable, and finally inquisitiveness and curiosity topics of the unit.*

RATIONALE

To analyze biological processes at the reductionistic level. Proteins- structure and function. Hierarchy in protein structures. Primary, secondary, tertiary, and quaternary structures. Proteins as enzymes, transporters, receptors, and structural elements. Engineering students need to understand the significance of reductionism in learning complex systems. Learning this will help to apply engineering principles for analyzing biological systems and develop energy-saving technologies that work in harmony with biosystems.

PREREQUISITES

Biology: Biomolecules (Class XI)

Chemistry: Biomolecules (Class XII)

UNIT OUTCOMES

List of outcomes of this unit is as follows:

U7-O1: Explain the significance of reductionism in understanding biosystems.

U7-O2: Describe various analytical techniques to estimate protein.

U7-O3: Deal with structures of protein and the hierarchy in four-level distributions of protein structures.

U7-O4: Discuss multiple functions of protein throughout the body.

Unit-7 Outcomes	EXPECTED MAPPING WITH COURSE OUTCOMES (1- Weak Correlation; 2- Medium correlation; 3- Strong Correlation)				
	CO-1	CO-2	CO-3	CO-4	CO-5
U7-O1	2	3	1	1	1
U7-O2	1	3	1	1	1
U7-O3	1	3	1	2	2
U7-O4	1	3	3	1	1

7.1 REDUCTIONIST APPROACH

The organization of life and its depiction in biology allow us to comprehend and hypothesize about the hierarchy of life on earth. The essential concern of the reductionist approach is whether the traits, theories, explanations, or techniques of one scientific domain (often at higher organizational levels) can be deduced from or explained by those of another scientific domain. In this chapter, we seek to get a reductionistic understanding of proteins and describe the various tasks that can be carried out as a result of chemical and physical interactions at the molecular level.

7.2 ANALYSIS OF PROTEINS

Protein analysis techniques include a diverse range of methods for detecting, purifying, and identifying proteins based on properties of the protein, like pH, binding affinity, size, and molecular weight. Chromatography and Electrophoresis techniques are used for the analysis of proteins.

7.2.1 Chromatographic techniques

Cation exchange: Ion-exchange chromatography exploits differences in the direction and magnitude of proteins' net electric charges at a certain pH. The column matrix is a synthetic polymer made of coupled with charged groups. The polymers containing anionic groups are referred to as cation exchangers, while those containing cationic groups are known as anion exchangers. This image illustrates ion-exchange chromatography on an anion exchanger. The pH and concentration of salt ions influence the binding affinity of each protein for the charged groups on the column. The separation can be increased by gradually adjusting the mobile phase's pH and/or salt concentration to establish a pH or salt gradient (Fig. 7.1).

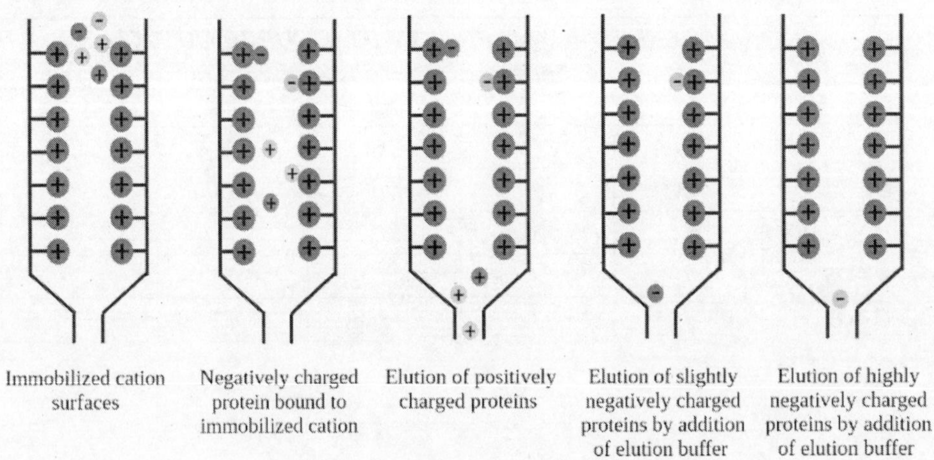

Fig. 7. 1 Anion exchange chromatography

(Images created using BioRender® software)

Size exclusion: The size attribute of molecules is used by size exclusion chromatography (SEC) to separate proteins through a column. The gel is made up of sphere-shaped beads with pores that vary in size. Since large protein molecules cannot pass through the pores, they are eluted into the column's void volume, whereas small molecules diffuse into the pores, and their flow through the column is decelerated due to their size. As a result, as molecules go through the column, they segregate based on their sizes and are eluted in decreasing order of molecular weight (Fig. 7.2).

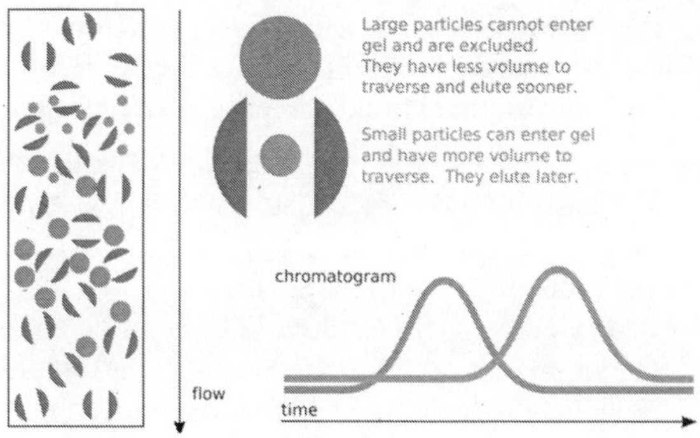

Fig. 7. 2 Size Exclusion Chromatography

Source: Wikimedia. Org (Creative Common License)

Affinity: Affinity chromatography is a separation technique that depends on a specific binding interaction between an immobilised ligand and its binding partner. The retained proteins bind precisely to the ligand affixed to the column. The proteins that do not bind to the column are eluted first, then the proteins that do bind to the column. Using a solution containing a competing free ligand or by modifying the pH, ionic strength, or polarity of the column, the bound protein is eluted from the column. (Fig. 7.3).

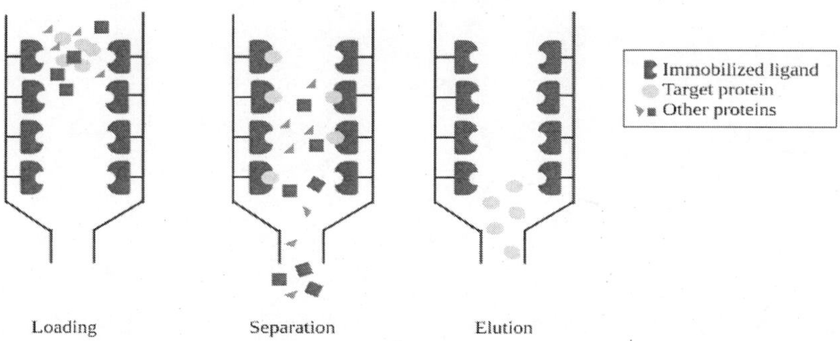

Fig. 7. 3 Affinity chromatography

(Images created using BioRender® software)

7.2.2 Electrophoretic techniques

SDS: Proteins can be divided using SDS-PAGE according to their molecular weight. SDS (sodium dodecyl sulphate) is an anionic detergent that binds proteins and masks them in a negative charge. One SDS molecule typically bonds to two amino acids. SDS imparts a negatiye charge to the protein, which replaces any intrinsic charge that the protein may have possessed earlier. Therefore, various proteins will have similar charge-to-mass ratios after exposure. Polyacrylamide gel electrophoresis is referred to as PAGE. During the process, SDS and ß-mercaptoethanol are added to the protein, followed by heating. SDS and ß -mercaptoethanol are to impart a negative charge to protein structure and break disulfide bonds, respectively. A polyacrylamide gel with the denatured protein mixture is added to and subjected to an electric field, which causes the proteins to travel in the direction of the positive electrode (the anode).

Isoelectric focusing:This method separates proteins according to its isoelectric points. A steady pH gradient is maintained in the gel by adding adequate ampholytes. A mixture of proteins is placed in a well on the gel. As soon as the electric field is applied, proteins enter the gel and move until it reaches a pH equivalent to its pI. Once proteins reach this pH, they stop migrating because pI is the pH at which they carry zero charges and will not be affected by an applied electric field. Isoelectric focusing includes the separation of proteins using non-denaturing gels in order to assess the biological activity of the separated proteins. The differentiation is somewhat based on net charge and partly on size. The employed gels are commercially available with an established pH gradient. These compounds are known as ampholines. When exposed to an electric field, the proteins migrate to the oppositely charged electrode, but they become immobile when they reach the isoelectric point of the protein in the gel. At low pH, the carboxylic acid groups of proteins tend to be uncharged and the amino groups positively charged, resulting in an overall positive charge for the proteins. At a high pH, the carboxylic acid groups of proteins tend to be negatively charged while the amino groups are uncharged, resulting in a positive total charge. A protein's isoelectric point is the pH at which it has no overall or net charge.

7.2.3 Protein Assays

Table 7. 1 Assays used for protein estimation

Assay method	Principle	Reagents
UV absorption	The UV absorbance at 280 nm can be used to determine the protein concentration because Tryptophan and Tyrosine amino acids absorb UV light at 280nm (commonly referred to as A_{280}).	
Protein-Cu complex → Chelation of protein with copper → Measurement of cuprous ion.	The peptide bonds in protein reduce Cu^{2+} (cupric) to Cu^{+} (cuprous) under alkaline conditions.	a) BCA (Bicinchoninic Acid): BCA forms a purple-colored complexwith Cu^{+} & exhibits a linear absorbance at 562 nm. The intensity of color formed is directly proportional to protein concentration. b) Folin-Ciocalteu Lowry: Folin-Ciocalteureagent (phosphomolybdic/ phosphotungstic acid) forms a complex with the cuprous ions and the side chains of tyrosine, tryptophan, and cysteine to produce a blue-green color that can be detected between 650 nm and 750 nm.
Protein-dye complex → Formation of Protein-dye complex → Measurement of change in color of complex	The dye-binding protein assay exploits the property of protein to bind to Coomassie dye under acidic conditions. The binding of protein to the dye causes a spectral shift from brown (Amax = 465nm) to blue (Amax = 610nm). The change in color intensity is directly proportional to protein concentration.	Coomasie dye (Bradford assay) The basic amino acids arginine, lysine, and histidine form a colored complex with dye.

7.3 PROPERTIES OF PROTEINS

Proteins are macromolecular polypeptides that are present in all living cells. They are present in every part of the body and serve as the basis for all of life's structures and functions. The term protein in Greek signifies fundamental or principal significance. Proteins are so named because they are significant chemical compounds necessary for the development and maintenance of all life. Animal or plant protoplasm contains 10–20% protein. Dairy products, including milk, grains, lentils, groundnuts, fish, and meat, are the primary sources of protein.

Proteins have the following essential properties:

a) Typically, proteins are generally colorless, odorless, and tasteless.

b) Proteins are large in size and therefore exhibit colloidal properties. They may significantly produce light-scattering in solution, thus resulting in visible turbidity which is called the Tyndall effect.

c) With the exception of the fundamental structure, all three levels of protein structure rely heavily on non-covalent connections, including hydrogen bonds, ionic interactions, hydrophobic interactions, and Van der Waal forces. Several things can affect non-covalent interactions such as heat, radiation, organic solvents, acids, bases, and salts. This disruption of native structure is termed denaturation. Both physical and chemical methods can be used to denature proteins. A protein's ability to retain its 3-D structure and function can be lost even in the presence of minor changes to its environment.

d) Proteins can function as an acid or a base depending on the pH of the environment since they include both basic (amino group) and acidic (carboxyl group) groups. Due to this characteristic, it is called to be amphoteric in nature.

The following are some crucial proteins that our body needs:

- Hormones: to coordinate numerous bodily processes,
- Enzymes: as biocatalysts for biochemical reactions,
- Antibodies: These help the bodies fight against infections and toxins.
- Transport proteins: To transport various substances in the blood to various body tissues,
- Structural proteins: Structural components of cell and tissue structures
- Contractile proteins: These help muscles and other cells contract. The elements carbon, hydrogen, oxygen, nitrogen, and sulphur are all controlled by proteins. Some may contain trace metals like iron, copper, zinc, manganese, etc.

All proteins can be partially hydrolyzed into peptides of different molecular weights, which can then be fully hydrolyzed to produce a-amino acids.

$$\text{Proteins} \dashrightarrow \text{Peptides} \dashrightarrow \alpha - \text{Amino acids}$$

An introduction to proteins and amino acids was provided in module 4. This lesson will cover how proteins are categorized according to their structures and functions.

PROTEINS

- **Classification on basis of Structure**
  - Primary Structure
  - Secondary Structure
  - Tertiary Structure
  - Quaternary Structure
- **Classification on basis of Functions**
  - Enzymes
  - Transport proteins
  - Receptor Proteins
  - Structural Proteins

7.4 STRUCTURE OF PROTEINS

A huge number of amino acids are joined together to form biopolymers called proteins through peptide bonds, which have three-dimensional (3D) structures. The organization of proteins is extremely intricate. A multisubunit protein's quaternary structure is an oligomeric structure in which different protein chains join together to create dimers, trimmers, tetramers, and other oligomers. The various chains in the oligomers may be the same proteins (homo-oligomers) combined together, or they may all be different proteins (hetero-oligomers). Noncovalent intermolecular forces or covalent disulphide bonds may be present in the oligomer to hold the various chains together.

7.4.1 Forces that stabilize protein structures

The forces that are present are as follows:

- Hydrogen bonding
- Anionic bonding
- Hydrophobic bonding
- Covalent bonding

Hydrogen bonding: These forces act between a hydrogen atom that is partially positive and a partially negative atom, such as O or N. There are two types of hydrogen bonding: Intramolecular and Intermolecular hydrogen bonding.

Anionic bonding: It is due to the interaction between cation and anion groups present in the side chain, which leads to the formation of the side chain linkages.

Hydrophobic bonding: Some amino acid side chains have hydrophobic linkages. In aqueous situations, proteins fold such that the hydrophilic polar side chains are distributed on the exterior or surface of proteins and the hydrophobic side chains are gathered within the folds.

Covalent bonding: The bond occurs between S atoms of two residues between two adjacent chains. Two polypeptide chains of 21 and 30 amino acid residues are linked by S-S cross-links to form the 51 amino acid long insulin.

7.4.2 Primary Structure:

The number and arrangement of amino acids make up a protein's basic structure. Amino acids are the monomers—or building blocks—of the polypeptide chain. Peptide bonds are the linkages between two protein monomers. The α-carboxyl group of one amino acid residue is bonded to the α-amino group of another through this bond. Therefore, the main chain of the polypeptide consists of central carbon attached to the carboxyl group, the amino group of amino acid, and R group forms the side chain.

The primary amino acid sequence is formed by a condensation reaction where the carboxylic acid of the upstream amino acid combines with the amine functional group of the downstream amino acid by eliminating a water molecule. The opposite reaction, known as hydrolysis, requires the addition of a water molecule to break the amide bond and separate the two amino acids.

The amide linkage between amino acids in protein structures is referred to as a peptide bond. Other amino acids will be added to the carboxylic acid terminal of the developing protein. As a result, proteins are always created in a specific direction, beginning with the amine, and finishing with the carboxylic acid tail. Never are new amino acids added to the amine of the first amino acid in the chain; instead, they are constantly added to the carboxylic acid tail. The ribosome controls the directionality of protein synthesis, also referred to as N- to C- synthesis.

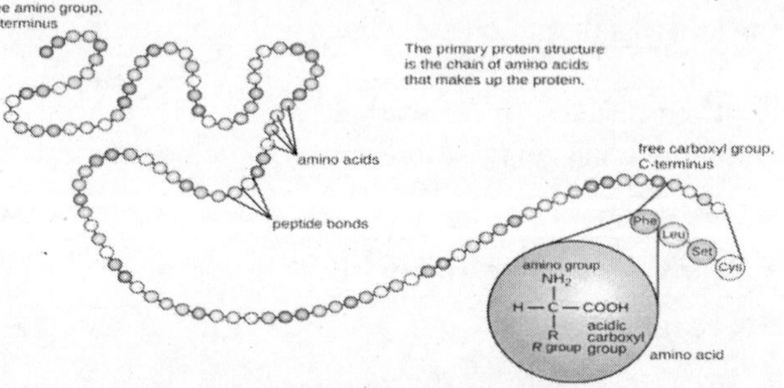

Fig. 7. 4 Primary protein structure

Source: wikimedia.org (Creative Common License)

7.4.3 Secondary structure

Secondary structure refers to the local patterns that develop due to interactions between amino acids that are proximal to one another during protein synthesis (polypeptides). The α-helix and β-strands are examples of these secondary structures. Types of hydrogen bonds distinguish these structures. Intramolecular hydrogen bonding leads to α-helix, and Intermolecular hydrogen bonding leads to β-strands/sheets.

α-helix: The polypeptide backbone develops a right-handed helical configuration in the α-helix. Right-handed refers to the direction of the turn in this context: The flexing fingers of the right hand depict the twists of the polypeptide as the thumb is pushed along the helix axis. It possesses a rod-like structure. The inner portion of the rod is made up of the main chain (the centre carbon, carboxyl group, and amino group of the amino acid), while the side chain (R group) stretches outward. Hydrogen bonding is what keeps the helical structure stable. It is formed in the main chain between the NH and CO groups. The NH group of each amino acid is hydrogen bonded to the CO group of the same amino acid, which is situated four residues ahead, i.e. it is formed between the carbonyl oxygen of ith amino acid and amino of $(i+4)^{th}$ amino acid. The α-helical structure results from intramolecular hydrogen bonding (Fig. 7.5). Helical structure can be right-handed or left-handed. Most of the known α- helices are right-handed. Globulin, Myosin, and Keratin are proteins that have α-helical structures.

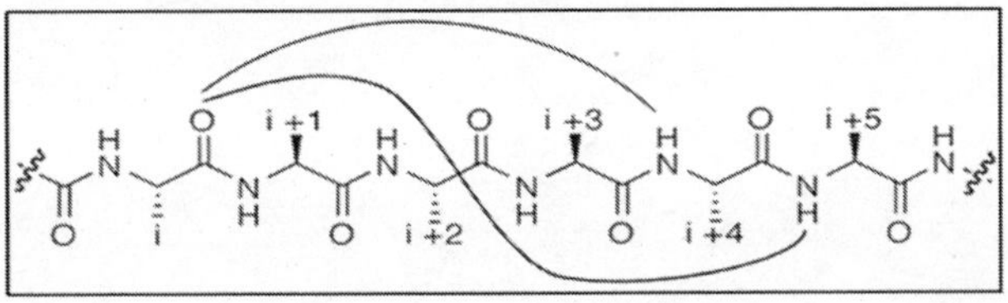

Fig. 7. 5 Intramolecular Hydrogen bonding in amino acid chain

(Source: Created with ChemDraw software)

β-pleated sheet: It is the second type of secondary structure. It is named β because this structure was elucidated after α- helix. The parallel alignment of polypeptide chains in a plane and the development of hydrogen bonds between the CO and NH of adjacent chains create β-pleated sheets. There are different kinds of sheets with pleats:

a) Parallel β-pleated sheet: The structure in which all of the polypeptide chains present have their N terminal on the same edge of the sheet and their C terminal on the opposite edges (Fig. 7.6).

b) Antiparallel β-pleated sheet: The structure is known as an antiparallel-pleated sheet if the alternating chains have their N-terminal ends on the same side. (Fig. 7.7).

The geometry of each amino acid molecule compels the hydrogen bonds to form at an angle that lengthens and weakens them, making the parallel arrangement less stable. In contrast, the hydrogen bonds in the antiparallel configuration are lined up exactly across from one another, resulting in more robust and stable connections.

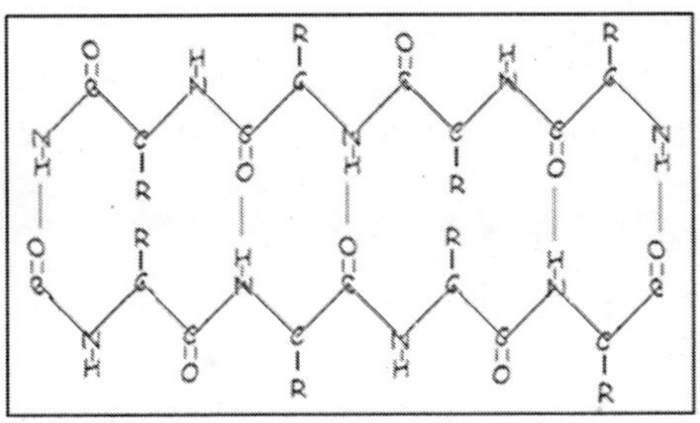

Fig. 7. 6 Parallel β-pleated sheet

(Source: Created with ChemDraw software)

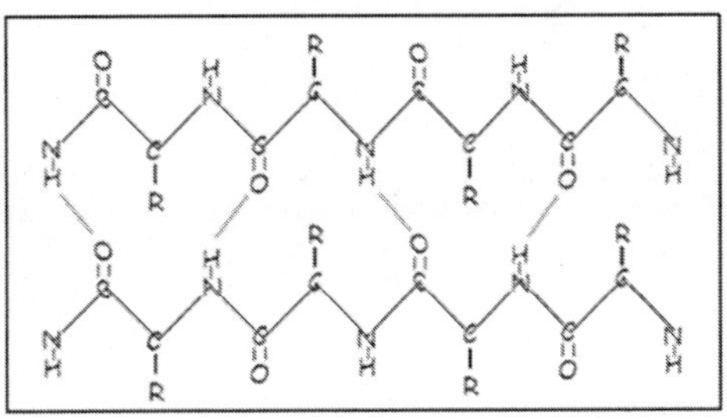

Fig. 7. 7 : Antiparallel β-pleated sheet

7.4.4 Tertiary Structure

Interactions between R groups mostly create the intricate three-dimensional tertiary structure of a protein. Multiple chemical interactions influence the tertiary structure of proteins. These include disulfide bonds, ionic bonds, hydrogen bonds, and hydrophobic interactions. Like-charged R groups repel one another, while unlike-charged R groups are drawn to one another (ionic bonds). Uncharged nonpolar side chains can form hydrophobic interactions. The interaction of cysteine side chains can create disulfide linkages. These strong and weak interactions create the ultimate three-dimensional form of the protein. A protein usually ceases to function when it loses its three-dimensional structure (Fig. 7.8).

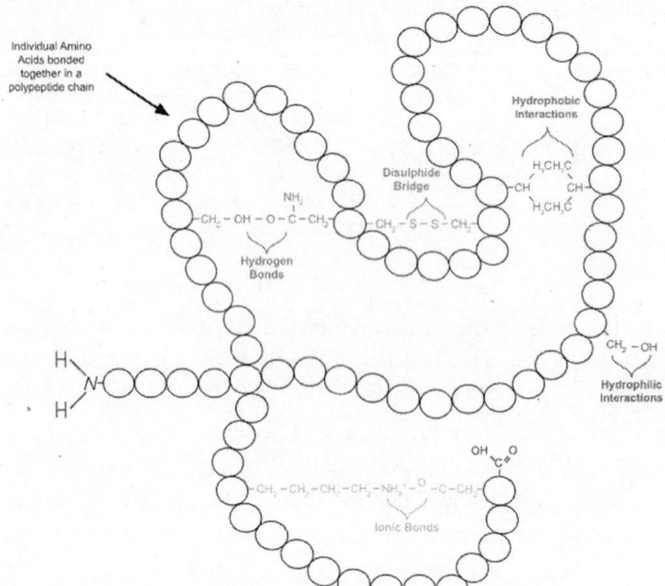

Fig. 7. 8 Tertiary structure of protein

Source: wikimedia.org (Creative Common License)

7.4.5 Quaternary Structure

A protein's quaternary structure is formed when several protein chains or subunits come together in a compact configuration. The primary, secondary, and tertiary structures are unique to each subunit. Hydrogen bonds and van der Waals forces between nonpolar side chains hold the subunits together.

For instance, consider the quaternary structure of hemoglobin. It comprises the two protein pairs known as the α- and β- chains. Every protein is attached to a heme molecule, commonly referred to as a prosthetic group. The secondary structures of myoglobin and the subunits of hemoglobin are virtually identical. Myoglobin and hemoglobin are parts of the globin family, a heme-containing globular polypeptides with eight α -helices in their protein fold. The helices of both proteins fit together similarly. The differences between hemoglobin and myoglobin are significant at the quaternary level. The two pairs of two types of closely related subunits, α and β chains, that make up hemoglobin form a tetramer. Since myoglobin is a monomer, it lacks a quaternary structure. Hemoglobin does not bind oxygen as firmly as myoglobin does. Myoglobin stores oxygen in muscle cells and releases it as needed, whereas hemoglobin absorbs oxygen from the lungs and delivers it throughout the body.

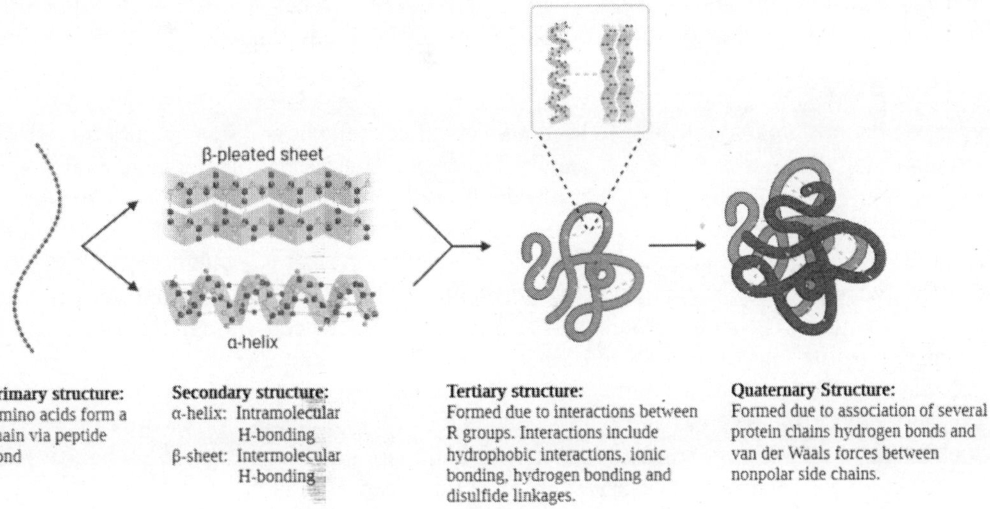

Fig. 7. 9 Protein packaging in eukaryotes

(Images created using BioRender® software)

7.5 FUNCTIONS OF PROTEINS

Proteins are dynamic molecules whose functions essentially depend on interactions with other molecules, and these interactions are influenced in physiologically significant ways by sometimes subtle and sometimes spectacular conformational changes in proteins. The significance of molecular interactions to the function of a protein cannot be overestimated.

Although the majority of these interactions are transitory, they may serve as the foundation for complicated physiological processes such as transport of biomolecules, immunological receptor function, and muscle contraction. Many protein functions include the reversible attachment of other molecules. These operations are carried out by proteins that exhibit the following essential characteristics of protein function. A ligand is a molecule that is irreversibly attached to a protein. A ligand may be any kind of molecule, including another protein. The transitory nature of protein-ligand interactions is vital to life because it enables rapid and reversible adaptation to changing environmental and metabolic settings. A ligand binds to a binding site on a protein that is in correspondance to the ligand in terms of charge, size, shape and hydrophobic / hydrophilic nature. In addition, the communication is precise, the protein can discriminate among the thousands of diverse molecules in its environment and bind to only one or a handful. A specific protein may contain unique binding sites for a variety of ligands. These precise molecular interactions are essential for preserving a biological system's high degree of order. Let us now look into different types of proteins based on their functions in their particular environment.

7.5.1 Transport Proteins

Most polar compounds are impermeable to lipid bilayers. Each cell has a plasma membrane, which aids in controlling the materials that enter and exit the cell. The plasma membrane contains a variety of membrane transport proteins, each of which is responsible for transferring a specific solute across the membrane. It enables the selective passage of certain water-soluble polar molecules from the surrounding environment. The plasma membrane has two main categories of transport proteins, which facilitate the passage of hydrophilic molecules. They are **Channel Proteins and Carrier Proteins**. A channel protein acts as a gateway into the cell through the membrane. These channel proteins introduce ions and other tiny molecules into the cell. Channel proteins are specific in nature, i.e., each channel protein can only accept one type of chemical. E.g., Calcium can only be transported into and out of cells using calcium channels. For different molecules, such as those for sodium, potassium, and chloride, there are specific types of channel proteins. To transport a particular solute across a membrane, carrier proteins bind it and then undergo conformational changes. In contrast, channel proteins have significantly weaker interactions with the solute being transported. They create aqueous pores that stretch across the lipid bilayer. When these pores are open, they permit solutes (often inorganic ions of the right size and charge) to pass through and therefore cross the membrane. Myoglobin and haemoglobin are possibly the most researched and best-understood proteins. For oxygen transport, multicellular organisms harness the characteristics of metals, most often iron due to its ability to sequester highly reactive oxygen to less reactive moeities. Transport mediated by channel proteins proceeds far more quickly than transport mediated by carrier proteins.

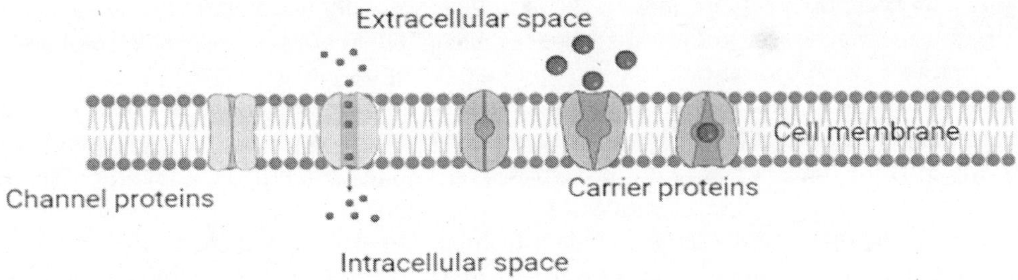

Fig. 7. 10 Transport Proteins: Channel & Carrier Proteins

(Images created using BioRender® software)

7.5.2 Receptor Proteins

A distinct class of proteins called receptors work by attaching to a particular ligand molecule. Receptors are divided into two categories: **Cell surface receptors and Intracellular receptors**.

Intracellular receptors are found inside the cell, majorly in the cytoplasm and nucleus. Ligands that bind to intracellular receptors are small and hydrophobic since they must be able to cross the plasma membrane to reach their receptors. E.g., Receptors for steroid hormones like estradiol and testosterone are intracellular. Intracellular receptors, which include those for steroid hormones, lipophilic vitamins, and chemical moieties like nitric oxide and hydrogen peroxide, need membrane-permeable ligands. Members of the family of steroid hormone receptors show similarities in chemical mechanisms and structural characteristics. The receptor is kept in the cytoplasm in the absence of a ligand by interacting with heat shock protein 90 (HSP 90), which masks the receptor's nuclear localization sequence. Once the ligand and receptor are combined, the HSP90 separates from the receptor, exposing DNA-binding sites. The receptor then rapidly migrates to the nucleus. Steroid receptors act as transcription factors; therefore, in the nucleus, they tend to bind to specific steroid response elements (specific binding sites along the chromosomal DNA) on the DNA. This either increases or decreases the transcriptional activity of the gene.

Surface receptor proteins are found on the membrane surrounding any cellular organelle, such as the cell membrane, nucleus membrane, etc. They can attach to appropriate ligands and initiate cellular signaling cascades. The receptor can change conformation when a ligand binds, sending a signal into the cell. The immune response is a combination of two systems that complement one another: the humoral and cellular immune systems. The humoral immune system is geared against bacterial infections and extracellular viruses (those found in body fluids), but can also respond to foreign proteins. The cellular immune system eliminates infected host cells as well as some parasites and alien tissues. At the core, humoral immune response are soluble proteins known as **antibodies or immunoglobulins**, abbreviated Ig. Immunoglobulins target for destruction bacteria, viruses, or big molecules that have been detected as foreign.

B lymphocytes, also known as B cells, generate immunoglobulins, which account for 20% of blood protein. At the centre of the cellular immune response is a subgroup of T lymphocytes called cytotoxic T cells (also known as killer T cells). T-cell receptors are proteins found on the surface of TC cells that identify infected cells or parasites. They identify and bind exogenous ligands, initiating cellular changes. These also belong to the category of defence proteins.

Transmembrane receptor proteins are embedded in the phospholipid bilayer. The hydrophobic part of the transmembrane receptor protein spans the phospholipid bilayer, while hydrophilic sections of the protein reach out on both the intracellular (cytoplasmic) and extracellular sides of the membrane. Typically, they can be split into three domains: extracellular, transmembrane, and intracellular. A receptor protein's extracellular domain is connected to ligand binding. E.g., SARS-CoV-2 requires host receptor protein AEC2 to bind. It is primarily found in the kidney, heart, intestine, lung, and lung. The SARS-CoV-2 spike (S) protein, which consists of the S1 and S2 subunits, is essential for receptor identification and cell membrane fusion. Angiotensin-converting enzyme 2 is the host receptor recognized and bound by the receptor-binding domain S1 subunit of the spike protein.

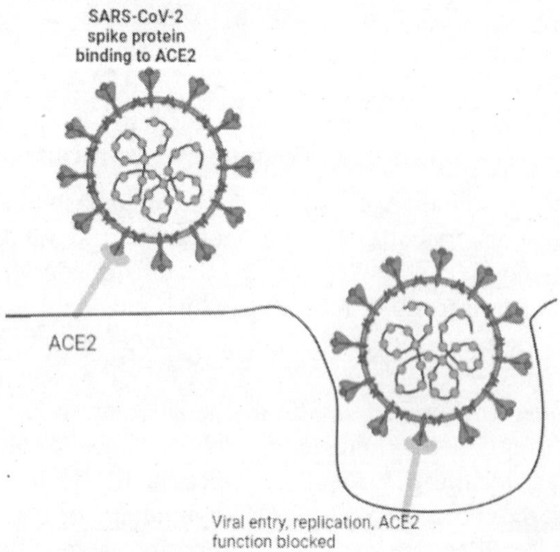

Fig. 7. 11 AEC2, a receptor protein to which spike protein of SARS-CoV-2 binds

(Images created using BioRender® software)

7.5.3 Hormonal Proteins:

In mammals, the neuroendocrine system is responsible for coordinating metabolism. When the organism's environment shifts, cells in one tissue respond by releasing a chemical messenger that travels to another cell in the same or a different tissue, where it binds to a receptor molecule and triggers a response. The chemical messenger (neurotransmitter; for example, acetylcholine) may only travel a fraction of a millimetre across the synaptic cleft to the next neuron in a network during neural signalling.

Hormonal signalling involves the delivery of hormones from the circulatory system to nearby cells or to distant organs and tissues, a distance that can exceed one metre. Despite this structural difference, the molecular signalling routes between the two are highly similar. Epinephrine and norepinephrine, for example, serve as neurotransmitters in certain synapses of the brain and smooth muscle, as well as hormones that regulate fuel consumption in the liver and muscle.

All hormones exert their effects on hormone-sensitive target cells via highly specialized receptors to which they bind with high affinity. Each cell type has a distinct combination of hormone receptors that defines its sensitivity to hormones. Furthermore, the intracellular targets of hormone activity may vary amongst cell types that express the same receptor, leading to varying responses to the same hormone. Because of the specificity in their interaction, hormones that are chemically similar but have different actions might be highly different. The interaction's high affinity enables cells to respond to extremely low hormone concentrations.

7.5.4 Enzymes

Enzymes are biological catalysts (referred to as biocatalysts) that accelerate biochemical reactions in living things. The enzyme is continuously employed during the reaction and is not destroyed. Each enzyme molecule found in a cell is unique and specific to a particular chemical reaction. Enzymes lower the activation energy of processes by binding to the reactants (substrates), bending the molecules, and disrupting their electron configurations. As a result, the molecules become unstable and reactive. The enzyme's active site, also known as the substrate binding site, is where the substrate binds to the enzyme. Enzymes are discussed in detail in chapter 5.

7.5.5 Structural Proteins

The structural protein family performs a wide range of functions, from controlling cell shape and mobility to supporting important structures like bones, cartilage, hair, and muscles. These proteins include collagen, myosin, keratin, and actin. Collagen, the most prevalent animal protein, is distributed throughout the body. It provides tensile strength and elasticity to connective tissue, including skin, ligaments, and tendons. Collagen is essential for the strength and firmness of skin; when collagen levels fall, your skin is more prone to wrinkles, potential damage, and sagging (Fig. 7.14). Based on the positioning of these proteins in the body they have different functions, elastin of ligaments, collagen of tendons, cartilages, bone and connective tissue fall under **structural proteins**.

Two proteins, myosin and actin, collaborate to generate the muscle's contractile force, also known as **contractile protein**. These proteins are organized into filaments that undergo brief connections and move past one another in order to induce contraction. Muscle protein is mostly composed of actin and myosin, which together account for about 80% of the total.

Casein, a phosphorylated protein found in milk and cheese, is not considered a structural protein. However, casein has recently been researched and commercially produced as a structural material in biomaterials. In 1898, casein plastic was created in Germany. Casein is made into a substance by adding acid to milk. The precipitated casein transforms into a thermoplastic with an ivory-like appearance. It is used industrially for impressions, clothing buttons, etc. There are similar protein from nature in seeds and the albumin from egg fall under **storage proteins**.

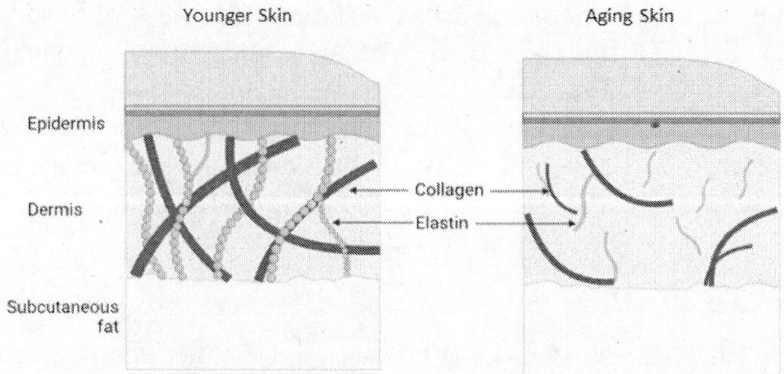

Fig. 7. 12 Collagen as structural protein of skin

(Images created using BioRender® software)

UNIT SUMMARY:

According to the reductionist theory, understanding a system's simpler components is crucial to know the system. One can think of this as a bottom-up method in biology, where it can be started at the most basic level and go way up to the most intricate system, where minute portions make up each new level of the whole.

Proteins are separated and purified based on differences in their properties such as size, charge to mass ratio, pH, binding affinities, and other qualities. Typically, electrophoresis and chromatographic methods are used for protein analysis. Chromatography includes ion exchange, size exclusion, and affinity chromatography, whereas electrophoresis includes SDS gel electrophoresis and isoelectric focusing.

The hierarchy of protein's structure includes four levels. The linear arrangement of amino acids is represented by the main structure. The secondary structure of a polypeptide chain is its spatial arrangement and twisting. The tertiary structure is the three-dimensional representation of the protein. Quaternary structure is the arrangement of identical or distinct polypeptide subunits.

Proteins are macromolecules that perform diverse functions, including transporting molecules throughout the body, acting as receptors for drugs and other molecules, maintaining the structural integrity of cells, and accelerating various biochemical reactions.

EXERCISES

Multiple Choice Questions

1) Myoglobin and Hemoglobin are ______ & ______, respectively.
   A. Primary & Secondary protein
   B. Secondary & Tertiary protein
   C. Tertiary & Quaternary protein
   D. Quaternary protein

2) Tertiary structure is maintained by
   A. Disulphide bond
   B. Peptide bond
   C. Hydrogen bond
   D. All of the above

3) Enzymes are
   A. Proteins
   B. DNA
   C. Carbohydrates
   D. Nucleic acids

4) The secondary structure is primarily maintained by
   A. Van der Waal force
   B. Hydrogen bond
   C. Ionic bonds
   D. Hydrophobic bonds

5) In α-helix, the bond is in between
   A. Amino acids close together
   B. Carbonyl oxygen of one amino acid to NH of 4th amino acid.
   C. Carbonyl oxygen of one amino acid to NH of 3rd amino acid.
   D. Carbonyl oxygen of one amino acid to NH of 5th amino acid.

6) Which statements are true regarding the quaternary structure of proteins?
   A. Refers to the organization and spatial arrangements of amino acids within a polypeptide chain
   B. Refers to the organization and spatial arrangements of proteins with many polypeptide chains
   C. Both a and b
   D. None of these

7) According to the Lock and Key model, the substrate acts as a
   A. Key
   B. Lock
   C. Inhibitor
   D. Enzyme

8) In the case of an induced fit model
   A. the substrate changes its shape slightly
   B. the enzyme changes its shape slightly
   C. none of them changes shape
   D. both changes shapes

9) Which of the following term-definition pairs are correctly matched?
   A. Product: the molecule that binds with an enzyme to start a reaction
   B. Active site: the small section of an enzyme where binding takes place
   C. Substrate: the molecule released at the end of a reaction
   D. Activation energy: the energy required to start enzyme binding

10) Which of the following statements about SDS polyacrylamide gel electrophoresis is correct?
   A. Wanted proteins can be tested for their biological activity after separation by SDS polyacrylamide gel electrophoresis.
   B. Proteins are solubilized but not denatured when separated by SDS polyacrylamide gel electrophoresis.
   C. SDS polyacrylamide gel electrophoresis separates proteins on the basis of charge.
   D. SDS polyacrylamide gel electrophoresis separates proteins on the basis of size.

11) Which of the following statements about isoelectric focusing is correct?
   A. Proteins separated by isoelectric focusing cannot be tested for biological activity.
   B. Proteins separated by isoelectric focusing can be tested for biological activity.
   C. The separation of proteins by isoelectric focusing is only based on charge.
   D. The separation of proteins by isoelectric focusing is only based on size.

12) The proteins are synthesized at
   A. Centrosomes
   B. Ribosomes
   C. Golgi bodies
   D. Mitochondria

13) Which proteins are called messenger proteins?
   A. Enzymes
   B. Hormones
   C. Storage
   D. Antibodies

14) Which of the following claims regarding column chromatography is false?
   A. Affinity chromatography involves the addition of groups or molecules that bind particularly to the desired protein to the column matrix.
   B. In reverse phase chromatography, the desired protein can be eluted selectively by solutions of varying hydrophobicity or ionic strength.
   C. Ion-exchange chromatography utilises several ionic groups linked to the column matrix that bind particularly to the desired protein.
   D. Proteins are separated by gel-filtration chromatography based on their capacity to bind to particular groups on the column matrix.

15) What are the membrane structures that function in transport?
   A. Channel proteins
   B. Carbohydrates
   C. Defensive Proteins
   D. Cholesterol

16) All of the following are found in membranes except:
   A. Nucleic acids
   B. Phospholipids
   C. Glycoproteins
   D. Glycolipids.

17) By adding SDS during the electrophoresis of proteins, it is possible to:
   A. Determine a protein's isoelectric point.
   B. Determine an enzyme's specific activity
   C. Determine the amino acid composition of the protein
   D. Separate proteins exclusively based on molecular weight.

18) In isoelectric focusing, the separation of proteins is based on
   A. Relative content of positively charged groups
   B. Relative content of negatively charged groups
   C. Both a and b
   D. pH

19) Which of the following methods could be used to check the molecular weight of your purified protein?
   A. SDS-PAGE
   B. Mass spectrometry
   C. Analytical SEC
   D. All of these

Answers: 1)C; 2)D; 3)A; 4)B; 5)B; 6)B; 7)B; 8)B; 9)B; 10)B; 11)B; 12)B; 13)B; 14)A; 15)A; 16)A; 17)A; 18)D; 19)D

Short Answer Type Questions

1) Explain the forces that stabilize protein structures.

2) Elaborate on different types of secondary protein structures.

3) What are the forces that stabilize the tertiary structure of a protein?

4) What is the primary difference between the Lock & Key and Induced Fit models?

5) How is carrier protein different from channel protein?

Long Answer Type Questions

1) Explain the hierarchy of protein structure.

2) Explain the two different mechanisms in which the enzyme functions.

3) Give a classification of enzymes based on nomenclature.

4) What are the various ways of quantifying proteins?

5) Explain the different functions of proteins.

6) Discuss various analytical techniques to estimate protein.

REFERENCES AND SUGGESTED READINGS

1. Hames, D., & Hooper, N., (Third Edition). (2005). *Biochemistry*. Taylor & Francis Group.
2. Nelson, D. L., & Cox, M. M., (Sixth Edition). (2013). *Lehninger Principles of Biochemistry*. Macmillan Learning.
3. Satyanarayana, U., & Chakrapani, U., (Fourth Edition). (2013). *Biochemistry*. Elsevier.
4. Jain, J. L., Jain, N., Jain, & Sunjay., (2005). *Fundamentals of Biochemistry*. S. Chand Publishing.
5. Khalsa, N., (2008). *Essentials of Biochemistry*. Aavishkar Publishers.
6. Wilson, K., & Walker, J., (2010). (Seventh Edition). *Principles and Techniques of Biochemistry and Molecular Biology*. Cambridge University Press.

Dynamic QR code for further reading

8 Metabolism

UNIT SPECIFICS

Through this unit we have discussed the following aspects:

- *Thermodynamics as applied to biological systems; Endergonic and exergonic reactions*
- *Free energy and biological reactions*
- *Redox potentials*
- *ATP as an energy currency*
- *Metabolic pathways - Catabolic, Anabolic, and Amphibolic reactions*
- *Integration of metabolic pathways in a cell and its physiological implications*
- *Synthesis of glucose from* CO_2 *and* H_2O *(Photosynthesis)*
- *Oxidation of Glucose (respiration) for energy*

The ***practical applications of the topics are discussed*** *for generating further curiosity and creativity as well as improving problem solving capacity.*

Besides giving a large number ***of multiple choice questions as well as questions of short and long answer types*** *marked in two categories following lower and higher order of* ***Bloom's taxonomy****, assignments through a number of numerical problems, a list of references and suggested readings are given in the unit so that one can go through them for practice. It is important to note that for getting more information on various topics of interest some* ***QR codes*** *have been provided in different sections which can be scanned for relevant supportive knowledge.*

After the related practical, based on the content, there is a ***"Know More"*** *section. This section has been carefully designed so that the supplementary information provided in this part becomes beneficial for the users of the book. This section mainly highlights the initial activity, examples of some interesting facts, analogy, history of the development of the subject focusing the salient observations and finding, timelines starting from the development of the concerned topics up to the recent time, applications of the subject matter for our day-to-day real life or/and industrial applications on variety of aspects, case study related to environmental, sustainability, social and ethical issues whichever applicable, and finally inquisitiveness and curiosity topics of the unit.*

RATIONALE

This chapter aims to teach students how chemistry is involved in biological processes. We came across thermodynamics in the 11th and 12th classes in chemistry, where we learned about various terms such as open system, closed system, isolated system, enthalpy, entropy, thermodynamics laws, etc. Now, here we will see how thermodynamics relates to the biological system. To perform useful work in our body, we need free energy. This energy is stored in ATP in the form of potential energy. We will learn how this energy will be available to perform each reaction in our cell. We do simple activities such as eating food, drinking water, inhaling, and exhaling without imagining how these processes work in our bodies. How the food we eat gets digested, and the nutrient is supplied to every cell of our body. Everyone knows that the oxygen we breathe and sustain our life on earth is a by-product of plants. This oxygen is freely available since plants perform photosynthesis to fix carbon dioxide (so next time, take good care of plants). We will explore how plants and some organisms in the presence or absence of light carry our essential reaction. Metabolism is a vast term that comprises catabolism and anabolism, through which our complex food is broken down into simpler molecules that can be absorbed by the cell. We will learn about all these pathways in detail in the following chapter.

PREREQUISITES

Chemistry- energy, thermodynamics, chemical reaction (Class XII)

Biology- cells, organelles of the cell (class X, XI)

UNIT OUTCOMES

List of outcomes of this unit is as follows:

U8-O1: Thermodynamics in biological systems and different types of reaction that follow thermodynamic law

U8-O2: Free energy and biological reactions, energy charge, equilibrium constant, endergonic and exergonic reactions

U8-O3: Metabolism- catabolic and anabolic pathways

U8-O4: Photosynthesis and Respiration

Unit-8 Outcomes	EXPECTED MAPPING WITH COURSE OUTCOMES (1- Weak Correlation; 2- Medium correlation; 3- Strong Correlation)				
	CO-1	CO-2	CO-3	CO-4	CO-5
U8-O1	-	-	-	1	3
U8-O2	-	-	-	1	3
U8-O3	1	1	-	-	3
U8-O4	-	-	-	-	3

8.1 INTRODUCTION

All living organisms need the energy to survive. They serve as energy transformers, absorbing and transforming energy into different forms. The energy transitions study in matter collection is referred to as thermodynamics. Cells absorb, convert, and release energy to carry out energetically unfavorable reactions. The rules of thermodynamics offer crucial insights into the reactions that take place in organisms and their bioenergetics. This chapter begins with a consideration of why living organisms require a steady energy supply. Then, we explore how the production and decomposition of various molecules generate energy for vital physiological activities. We will also learn about the metabolism and multiple pathways involved in energy generation and consumption. And lastly, we will compare cellular respiration and photosynthesis.

8.2 THERMODYNAMICS: APPLIED TO BIOLOGICAL SYSTEMS

Energy can be transformed by cells from one form to another. The first law of thermodynamics or energy conservation states that energy cannot be generated or destroyed but can be changed from one form to another. There must always be the same amount of energy in the universe altogether. As an animal cell breaks down food, for instance, some energy in the chemical bonds in the food molecules (chemical bond energy) is converted into the thermal motion of molecules (heat energy). However, if the reactions inside the cell result in the universe becoming more disordered—as required by the second law, as outlined below—then this conversion of chemical energy into thermal energy is crucial.According to the second law of thermodynamics, the universe is always moving toward higher degrees of disorder. When left unattended, non-living things gradually disintegrate into disorder. Living cells, on the other hand, create and preserve order on every level. This is possible because of complex molecular mechanisms that convert environmental energy into chemically bound energy. Even though the materials from which they are produced are continuously degraded, changed, and recycled, they retain their shape. Even while you now have atoms that were, for the most part, not in your body a decade ago, our body still has a similar basic structure as it had then. Living cells create order and might thus seem to violate the thermodynamics second law by surviving, developing, and building complex communities and even entire organisms. A cell is not an isolated system; hence this is not true. A cell uses energy from its surroundings, such as food, inorganic molecules, or photons of sunlight, to create order within itself by creating new chemical bonds and huge macromolecules. Heat is a by-product of useful chemical reactions carried out in the body. It is also produced when molecules participate in a random collision. The heat energy the cell's reactions produce is rapidly diffused into its surroundings since it is not an isolated system. There, the heat intensifies the thermal movements of adjacent molecules, increasing the entropy of the environment. The heat released by a cell must be sufficient to raise the disorder outside the cell's environment. Only in this instance is the second law of thermodynamics satisfied since the chemical reactions inside the cell increase the system's overall entropy.

8.2.1 Bioenergy and biological reactions

Biological Reactions: Living things need an ongoing supply of energy for three main reasons: 1) the execution of mechanical work in cellular and muscular contraction, 2) the active movement of ions and molecules, and 3) the synthesis of complex biomolecules from basic building blocks. The processes that keep an organism in a condition that is out of equilibrium consume free energy from the environment. Six biological reactions are mediated by enzymes in biological systems. These include isomerization, ligation, hydrolysis, oxidation-reduction, group transfer, and synthesizing and removing carbon-carbon double bonds. Each reaction is either energy-utilizing (endergonic) or an energy-releasing (exergonic) reaction. The difference between these two biological reactions is discussed in **Table 8.1.**

Table 8. 1 Difference between endergonic and exergonic reactions

Endergonic reactions	Exergonic reactions
vorable or nonspontaneous reactions	rable or spontaneous reactions
energy from the environment.	;fer energy into the environment
ıical bonds that are created as a result of the reaction are less strong than those that were broken.	ger chemical bonds than those in the reactants were broken throughout the reaction.
em's free energy increases	m's free energy decreases
lard Gibbs Free Energy (G) is positive (greater than 0)	lard Gibbs Free Energy (G) is negative (less than 0)
py (S) decreases	py (S) increases
ıples are photosynthesis, and the melting of ice into liquid water	ıples are cellular respiration, combustion, and chemiluminescence

8.2.2 Free-Energy Change, ΔG

Paper burns readily as,

paper + $O_2 \rightarrow$ smoke + ashes + heat + CO_2 + H_2O

At the same time, chemical bonds are broken to release energy during this process, and paper loses its structural integrity and chemical identity. In the language of thermodynamics, free energy is released to perform work or to drive chemical reactions. The path "downhill" along the energy curve is the spontaneous direction for any reaction. A "downhill" reaction is considered to be energetically beneficial in this sense.

Consider a simple reaction $Y \rightarrow X$,

$$\Delta G = \Delta G^{\circ} + RT \ln [X]/[Y]$$

Where

ΔG is in kilocalories per mole, [Y] and [X] denote the concentrations of Y and X in moles/liter, 'ln' is the natural logarithm, and RT is the product of the gas constant, R, and the absolute temperature, T. At 37°C, RT = 0.616 (A mole is 6.022×10^{23} molecules of a substance.)

Energetically beneficial reactions, or those with a negative ΔG, cause the disorder by reducing the system's free energy. Conversely, energetically unfavorable processes create order and have a positive ΔG. Life is made possible by the ability of enzymes to link energetically unfavorable reactions with energetically favorable ones, thereby establishing biological order. Because ΔG varies as products accumulate and substrates are depleted, chemical reactions frequently go on until equilibrium is reached. Because the speeds of the forward and reverse reactions are similar at equilibrium, there is no net change in the substrate or product concentrations. Such chemical inactivity does not support life. Living cells constantly exchange materials with their surroundings to replenish nutrients and remove waste, preventing the development of full chemical equilibrium.

Comparison of the energetics of several reactions is possible using the Standard Free-Energy Change, ΔG°.

We must look to the typical free-energy change of a reaction, ΔG°, to compare reactions. Some of the reactions with their ΔG° are listed in **Table 8.2.** In ideal conditions, where the concentrations of all the reactants are kept at the same fixed value of 1 mole/liter, the ΔG° is concentration independent and solely depends on the intrinsic properties of the interacting molecules.

When [X]/[Y] = 1, the value of ΔG equals the value of ΔG°.

This relationship demonstrates that the reaction's direction is exclusively decided by the properties of the molecules when reactants and products are present in equal proportions.

Table 8. 2 Standard free energies for different types of reactions

Reaction type	$\Delta G'^{\circ}$ kJ/mol	$\Delta G'^{\circ}$ kal/mol
Hydrolysis reaction		
Acid anhydrides		
Acetic anhydride → H_2O + 2acetate	-91.10	-21.80
ATP + H_2O → ADP + Pi	-30.50	-07.30
ATP + H_2O → AMP + PPi	-45.60	-10.30
PPi → H_2O + 2Pi	-19.20	-04.60
UDP-glucose + H_2O → UMP glucose 1 - phosphate	-43.00	-10.30
Rearrangements		
Glucose 1-phosphate → Glucose 6-phosphate	-07.30	-01.70
Fructose 1-phosphate → Glucose 6-phosphate	-01.70	-00.40
Elimination of water		
Malate → Fumarate + H_2O	03.10	00.80
Oxidations with molecular oxygen		
Glucose + $6O_2$ → $6CO_2$ + $6H_2O$	-2,840.00	-686.00

Source: Principles of Biochemistry (V Edition), By Nelson, D. L.; and Cox, M. M.W.H. Freeman, and Company

The Equilibrium Constant is Directly Proportional to ΔG°

For the simple reaction Y → X, the ratio of the substrate to the product at equilibrium is called the reaction's equilibrium constant, *K*. Expressed as an equation:

$$K=[X]/[Y]$$

Where [X] is the concentration of the product and [Y] is the concentration of the substrate at equilibrium.

$$\Delta G = \Delta G^{\circ} + RT\, ln[X]/[Y]$$

We can see that, at equilibrium at 37°C, where ΔG = 0 and the constant RT = 0.616, this equation becomes:

$$\Delta G^{\circ} = -0.616\, ln[X]/[Y]$$

In other words, ΔG° is directly proportional to the equilibrium constant, K:

$$\Delta G° = -0.616 \ln K$$

$$\Delta G° = -1.42 \log K$$

Concentrations of all reactants ***and products are included in the equilibrium constant in complex reactions***

For two reactants to combine & form a single product:

$$A + B \rightleftharpoons AB$$

$$K = [AB]/[A][B]$$

Concentrations of both reactants are multiplied because the formation of product AB depends on

The collision of A and B, and these encounters occur at a rate proportional to [A] × [B].

The Equilibrium Constant indicates the Strength of Molecular Interactions.

If the free-energy change for the interaction is negative, or if the combined free energies of the two molecules when they are unbound are lower than the free energy of the resulting complex, then two molecules will bind to one another. K is frequently used as a measure of the degree of non-covalent interaction between two molecules since the equilibrium constant of a process is directly proportional to ΔG°. The critical strength is useful because it shows how precisely the two molecules interact. As the binding energy (energy generated during the binding exchange) rises, K becomes increasingly essential. Consequently, the two molecules will bond more securely. The greater K is, the more significant the reduction in free energy between the dissociated and connected states. An impactful alteration in a binding interaction can result from even a small shift in non-covalent connections.

The Changes in Free Energy for Sequential Reactions Are Additive

With energetically favorable reactions, enzymes catalyze energetically unfavorable reactions. However, the pathway moves quickly toward completion because the total ΔG° for the series of sequential reactions has a significantly negative value. Other, broader strategies for coupling processes with the aid of enzymes include the creation of activated carriers such as ATP that can transfer energy from one reaction site to another.

8.2.3 Activated Carriers of Electrons

Activated carriers are molecules that can degrade and release free energy upon their degradation. The most common activated carriers are ATP, GTP, NADH, and NADPH, as discussed in **Table. 8.3.** The energy they release is used to power different chemical reactions.

Table 8. 3 Activated Carriers with their high-energy chemical groups

Activated Carrier	Group Carried in High-Energy Linkage
ATP	Phosphate
NADH, NADPH, $FADH_2$	Hydrogen and Electrons
Acetyl CoA	Acetyl
Carboxylated biotin	Carboxyl
S-adenosylmethionine	Methyl
Uridine diphosphate glucose	Glucose

Source: Bruce Alberts, Dennis Bray, Karen Hopkin, Alexander Johnson, Julian Lewis, Martin Raff, Keith Roberts, Peter Walter. Essential Cell Biology. Third Edition

ATP is the Most Widely Used Activated Carrier

ATP fuels all cellular activity. An endergonic process is driven by coupling an exergonic process with ATP. Adenosine 5'-triphosphate (ATP) is a useful and efficient kind of energy that can drive several chemical processes in cells. Adenosine 5'-diphosphate (ADP) is converted to ATP by adding phosphate in an energetically unfavorable phosphorylation process. To make ADP and inorganic phosphate (Pi), ATP hydrolyzes this energy packet when necessary. The newly generated ADP is used for yet another round of the phosphorylation procedure to produce ATP, the most common activated carrier in cells. This produces energy for several pumps that actively transfer materials into or out of the cell, establishing an ATP cycle in the cell (Fig. 8.1). Additionally, it powers the molecular motors that enable the contraction of muscle cells and the transport of materials along the long axons of nerve cells. It is still unknown, nevertheless, why this particular nucleotide was selected by evolution to act as the main energy carrier above the others. Despite similarities, the functions of the nucleotide GTP in cells are very different.

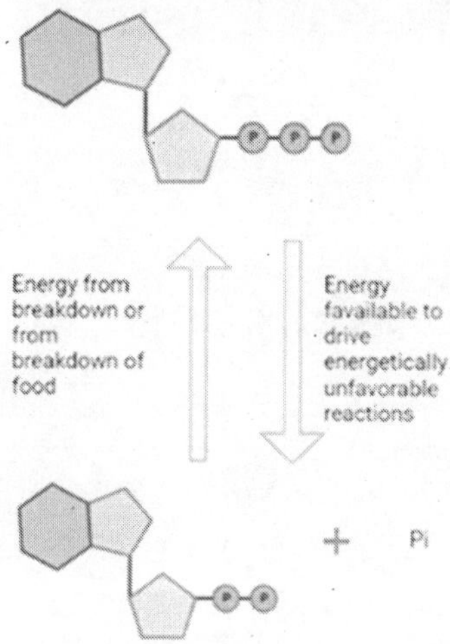

Fig. 8. 1 Interconversion of ATP and ADP occurs in a cycle

(Images created using BioRender® software)

Every minute, ATP in a cell is recycled. Every day, humans use around their body weight of ATP. No ATP generation results in an eventual demise. The products are more stable than the reactants in hydrolysis processes with substantial, negative, standard free-energy shifts for one or more of the following reasons: Charge separation, as in the case of ATP, relieves bond strain in reactants caused by electrostatic repulsion; (2) the products are stabilized by ionization, as in the case of ATP, acyl phosphates, and thioesters; (3) the products are stabilized by isomerization (tautomerization), as in the case of PEP; and/or (4) the products are stabilized by resonance, as in the case of creatine released from phosphocreatine.

8.2.4 Concept of Energy charge

The high level of ATP in the cell is necessary for metabolism. A measurement of the relative concentration of nucleotide phosphate in the cell helps assess how "charged" a cell is to carry out useful work. It compares the number of high-energy phosphoanhydride bonds that can be hydrolyzed to the amount of ATP, ADP, and AMP present in all cells (Fig. 8.2). The energy charge can be between zero (all AMP) and one (all ATP). The normal range for healthy cells is 0.80 to 0.95. The quantity of ATP available for metabolic processes is indicated by the Energy Charge (EC), a measurement of [ATP], [ADP], and [AMP] in the cell. While a high EC drives anabolic pathways, a low EC stimulates catabolic mechanisms, which restore ATP levels.

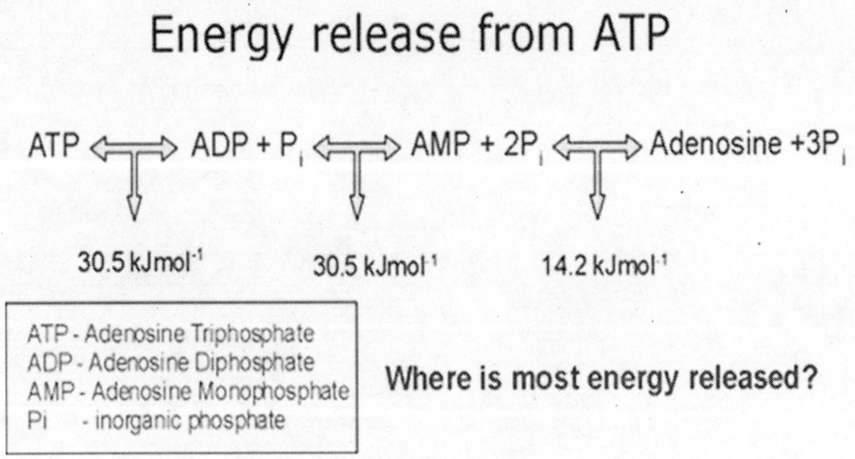

Fig. 8. 2 Energy release from ATP

(Images created using BioRender® software)

Reaction coupling: Two molecules are frequently joined together by the energy stored in ATP

Consider the following reaction,

A–H + B–OH → A–B + H2O

1. B–OH + ATP → B–O–PO3 + ADP

2. A–H + B–O–PO3 → A–B + Pi

Net result: B–OH + ATP + A–H → A–B + ADP + Pi.

The coupling to ATP hydrolysis in an enzyme-catalyzed reaction pathway has forced the energetically unfavorable condensation reaction. The energy that is transferred from catabolism to anabolism. Both endergonic and exergonic reactions are propelled by energy from exergonic processes. Ex. The cycle of cellular respiration in photosynthesis. By adding a phosphate group to another molecule, the exergonic hydrolysis of ATP and the endergonic dehydration process are connected.

8.2.5 Oxidation and Reduction Reactions

The plant and animal energy in food is released by slow oxidation or controlled combustion. Therefore, a cell can create bonds between C and O to obtain energy from sugars or other organic molecules. This process is known as cellular respiration.

Oxidation and Reduction Involve Electron Transfers Oxidation is introducing oxygen atoms to a molecule. In this context, the elimination of electrons from an atom is called oxidation. An atom gains electrons in the opposing reaction called reduction. Because chemical processes conserve the number of electrons, oxidation and reduction always occur in tandem (there is never a net gain or loss). When a molecule in a cell absorbs an electron (e-), a proton (H+) is typically picked up at the same time (protons being freely available in water). In this case, the end outcome is to add one extra hydrogen atom to the molecule:

$$A + e^- + H^+ \rightarrow AH$$

Even though a proton and an electron are involved (rather than just an electron), these hydrogenation reactions are reductions, whereas dehydrogenation processes are the opposite and are oxidations. An easy way to tell if an organic molecule is being reduced or oxidized is to count the C-H bonds; reduction happens when the number of C-H bonds increases, whereas oxidation occurs when the number of C-H bonds decreases.

8.3 METABOLIC PATHWAYS

Metabolism is the sum of the total chemical activities of all cells to convert food into energy. Cells must perform an unending series of chemical reactions to create the chemicals necessary for the organism to meet its metabolic needs (Fig. 8.3). It is possible to think of each cell as a little chemical factory that carries out millions of these reactions per second. Chemical reactions that only take place at temperatures significantly higher than those inside a cell are performed by biocatalysts. It requires a massive increase in chemical reactivity, provided by a specialized group of proteins called enzymes, which helps speed up every chemical reaction.

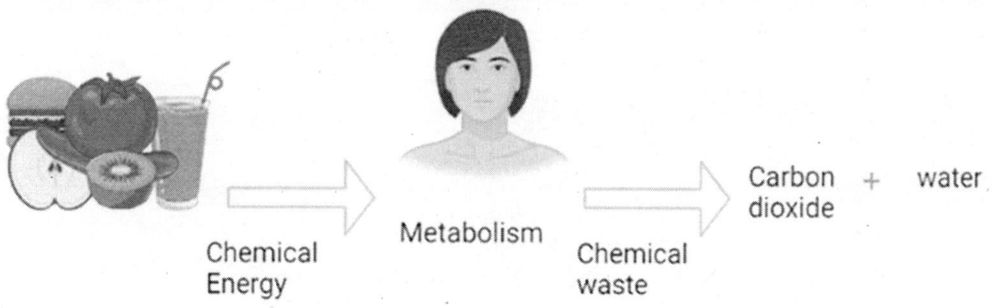

Fig. 8. 3 Flow of energy in metabolism

(Images created using BioRender® software)

The products of one reaction serve as the starting materials for subsequent ones in these enzyme-catalyzed processes, typically coupled in series. A complex network of interlinked reactions is created when the lengthy linear reaction routes, also known as metabolic pathways, are connected. Because it enables the cell to precisely control its metabolism, catalysis is advantageous. Each successive step in a metabolic route results in a modest, precise chemical change, typically the elimination, transfer, or addition of a single atom or functional group. A succession of metabolic intermediates called metabolites turns the precursor into the final product. Intermediary metabolism frequently refers to the sum of all metabolic pathways that convert precursors, metabolites, and low-molecular-weight products.

In **Catabolism,** the complex organic molecules such as carbohydrates, lipids, and proteins are broken down into smaller, simpler molecules such as lactic acid, CO_2, and NH_3. The remaining energy is dissipated as heat. Whereas **anabolism**, known as the biosynthesis phase, simple precursors are converted into larger, more complex compounds, including lipids, polysaccharides, proteins, and nucleic acids. Anabolic reactions require energy, often provided by the phosphoryl group of ATP and the reducing power of NADH, NADPH, and $FADH_2$ (Fig. 8.4).

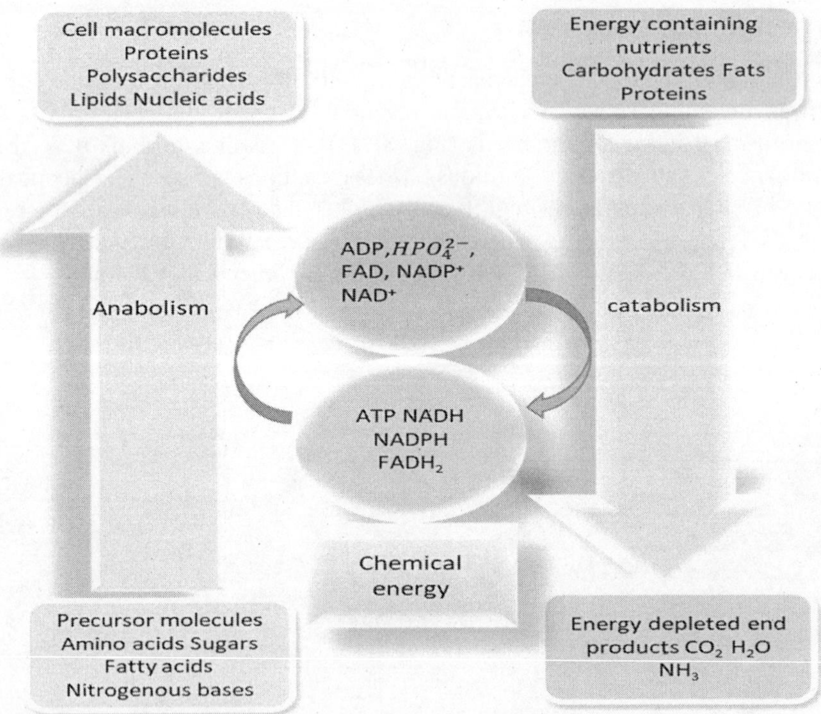

Fig. 8. 4 Catabolic and anabolic pathways and their energy relationship

8.3.1 The process of cellular respiration

Enzymes reduce complex organic compounds into simpler ones during the catabolic process (as explained above). During the initial phase of catabolism, enzymes reduce complex polymeric food molecules into simpler monomeric subunits. Digestion happens either outside of cells (in the intestine) or in specialized organelles within cells called lysosomes. During stage 2 of glycolysis, glucose molecules are divided into two smaller pyruvate molecules in the cytosol. The Krebs cycle in mitochondria constitutes the third step of catabolism. In the final step, the NADH and [$FADH_2$] generated during the glycolysis Krebs cycle are oxidized via the electron transport chain, which releases energy in the form of ATP. The ATPs are then utilized as the energy currency of the cell.

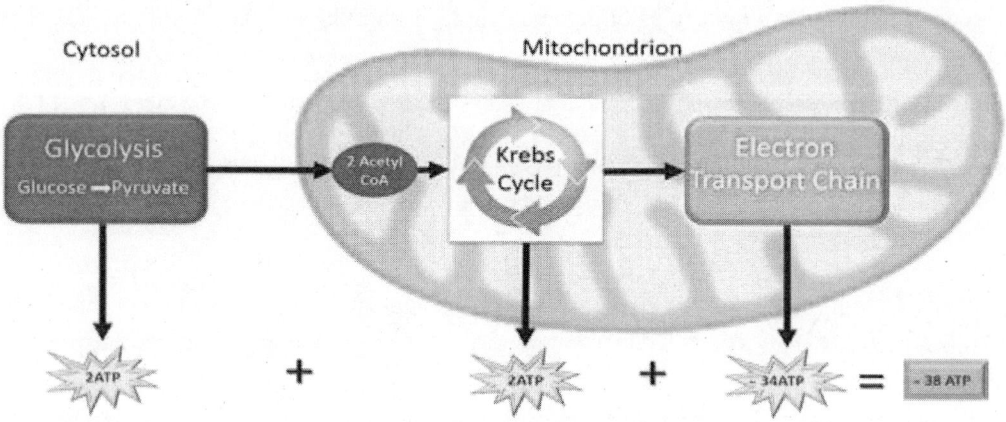

Fig. 8. 5 Overall cellular respiration

(Creative commons licences)

8.3.2 Glycolysis

In glycolysis (from the Greek glykys, "sweet" or "sugar," and lysis, "splitting"), a molecule of **Glucose** is broken down by a sequence of enzyme-catalyzed reactions into two molecules of the 3-carbon **pyruvate**. During the sequential steps of glycolysis, as shown in **Fig. 8.6**, some of the free energy generated from Glucose is preserved in the form of ATP and NADH. Glycolysis was the first recognized metabolic route and is likely the best understood. Also known as Embden-Meyerhof-Parnas (EMP pathway), which was discovered by **Gustav Embden, Otto Meyerhof, and Jakub Karol Parnas**. Glycolysis is the pathway with the largest carbon flux in most cells and a nearly universal primary mechanism of glucose catabolism. In certain mammalian tissues/cells such as erythrocytes, brain, renal medulla, and sperm, the glycolytic breakdown of Glucose is the only source of metabolic energy. Glycolysis occurs in aerobic and anaerobic organisms as this pathway does not require oxygen.

The six-carbon Glucose is converted into two molecules of the three-carbon pyruvate in a series of ten steps; the first five processes represent the preparation phase in which mainly the 6-C Glucose is phosphorylated and cleaved to yield 2 molecules of the 3-C compound.

1) In the first step, Glucose is phosphorylated at the hydroxyl group on the C-6 position with the help of an enzyme hexokinase along with cofactor magnesium ions.
2) Second step is the isomerization step; D-glucose 6-phosphate is converted to D-fructose 6-phosphate in the presence of phosphoglucose isomerase.
3) D-fructose 6-phosphate is phosphorylated using ATP, but this time at the C-1 position to yield D-fructose 1, 6-bisphosphate.
4) Fructose 1, 6-bisphosphate is then split into two three-carbon molecules, one is dihydroxyacetone phosphate (DAP) and the other glyceraldehyde 3-phosphate (GAP); the name "Glycolysis" of the pathways comes from this step, as there is **lysis** of **Glucose**.
5) The dihydroxyacetone phosphate (DAP) is isomerized to glyceraldehyde 3-phosphate (GAP), thereby forming two molecules of glyceraldehyde 3-phosphate at the end of the first phase of glycolysis.

It should be noted that, till now, two molecules of ATP have been invested for the cleavage of Glucose into 2 three-carbon pieces; this investment will yield a healthy return. In the preparation phase of glycolysis, ATP is used to increase the free energy content of the intermediates, and the carbon chains of all digested hexoses are transformed to glyceraldehyde 3-phosphate.

The gain in energy occurs during the payoff phase of glycolysis.

6) The two molecules of glyceraldehyde 3-phosphate are oxidized and phosphorylated by inorganic phosphate (Pi + NAD^+) to produce 1, 3-bisphosphoglycerate in the presence of glyceraldehyde 3-phosphate dehydrogenase.
7) In the next step, energy is released/gained in the form of ATP by the action of an enzyme phosphoglycerate kinase on 1, 3-bisphosphoglycerate to convert it into 3- phosphoglycerate.
8) There is a rearrangement of the phosphate group in this step, 3- phosphoglycerate is converted to 2- phosphoglycerate.
9) 2- phosphoglycerate is converted to Phosphoenolpyruvate catalyzed by enzyme enolase along with the magnesium ions as a co-factor.
10) In the final step, pyruvate is formed along with the gain of 2 molecules of ATP in the presence of the enzyme pyruvate kinase.

The paired phosphorylation of four ADP molecules to ATP conserves the majority of this energy. Due to the investment of two molecules of ATP during the preparation phase, the net yield is two molecules of ATP per molecule of Glucose. During the payoff phase, energy is also preserved by producing two molecules of the electron transporter NADH per glucose molecule.

In the sequential reactions of glycolysis, three types of chemical transformations stand out: (1) the degradation of the carbon skeleton of Glucose to produce pyruvate; (2) the phosphorylation of ADP to ATP by compounds with high phosphoryl group transfer potential that are formed during glycolysis; and (3) the transfer of a hydride ion to NAD to produce NADH.

Glycolysis and Glycolytic Enzymes

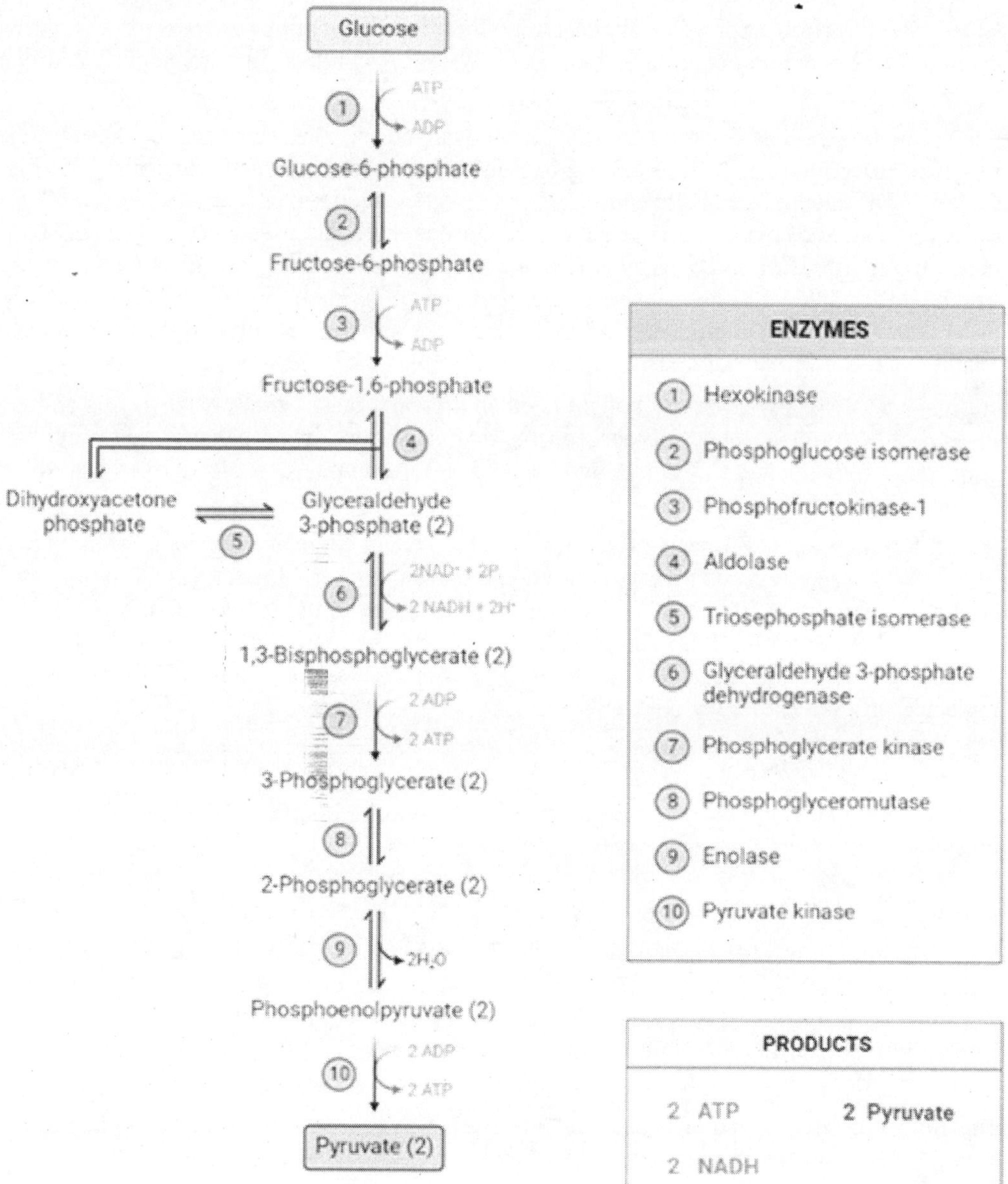

Fig. 8. 6 Reactions in Glycolysis

(Images created using BioRender® software)

Fates of Pyruvate - The pyruvate produced by glycolysis is subsequently digested via one of three catabolic processes, except for a few exceptions among bacteria. In aerobic organisms, glycolysis is the initial step in fully breaking down Glucose. Pyruvate is then oxidized to yield the acetyl group of acetyl-coenzyme A, with loss of CO_2; this acetyl group is then completely oxidized to CO_2 during the citric acid cycle. The electrons from these oxidations are passed to O_2 through a chain of carriers in mitochondria to form H_2O. The ATP synthesis in mitochondria is driven by the energy obtained from the electron transfer reactions.

The second route for pyruvate is its reduction to lactate via lactic acid fermentation. It takes place under low oxygen conditions, such as during exercise, the muscle contracts vigorously; during this stage, NADH cannot be reoxidized to NAD, but NAD is required as an electron acceptor for the further oxidation of pyruvate. Under these conditions, pyruvate is reduced to lactate, accepting electrons from NADH and thereby regenerating the NAD necessary for glycolysis to continue. Certain tissues and cell types (retina and erythrocytes, for example) convert Glucose to lactate even under aerobic conditions. Lactate is also the product of glycolysis under anaerobic conditions in some microorganisms.

The third major route of pyruvate catabolism is ethanol formation. In some plant tissues and certain invertebrates, protists, and microorganisms such as brewer's or baker's yeast, pyruvate is converted to ethanol and CO_2 under hypoxic or anaerobic conditions known as ethanol (alcohol) fermentation.

The Overall Balance Sheet Shows a Net Gain of ATP - Now, we can construct a balance sheet for glycolysis to account for (1) the fate of the carbon skeleton of Glucose, (2) the input of Pi and ADP and output of ATP, and (3) the pathway of electrons in the oxidation-reduction reactions. The left-hand side of the following equation shows all the inputs of ATP, NAD, ADP, and Pi, and the right-hand side shows all the outputs (keep note that each molecule of Glucose yields two molecules of pyruvate)

$$\text{Glucose} + 2\text{ATP} + 2\text{NAD} + 4\text{ADP} + 2\text{Pi} \rightarrow 2\text{ pyruvate} + 2\text{ADP} + 2\text{NADH} + 2\text{H}^+ + 4\text{ATP} + 2\text{H}_2\text{O}$$

We can balance the equation by removing common terms on both sides of the equation, the overall equation for glycolysis under aerobic conditions:

$$\text{Glucose} + 2\text{NAD} + 2\text{ADP} + 2\text{Pi} \rightarrow 2\text{ pyruvate} + 2\text{NADH} + 2\text{H}^+ + 2\text{ATP} + 2\text{H}_2\text{O}$$

These two molecules of NADH formed by glycolysis in the cytosol are, under **aerobic** conditions, **re-oxidized to NAD by transfer of their electrons to the electron transfer chain**, which in eukaryotic cells is located in the mitochondria. The electron-transfer chain passes these electrons to their ultimate destination, O_2: Electron transfer from NADH to O_2 in mitochondria provides the energy for synthesizing ATP by respiration-linked phosphorylation.

Oxidation of Pyruvate - Pyruvate decarboxylation or **pyruvate oxidation,** also called the **link reaction** (or **oxidative decarboxylation of pyruvate**), is the conversion of pyruvate into acetyl-CoA by the enzyme complex pyruvate dehydrogenase complex.

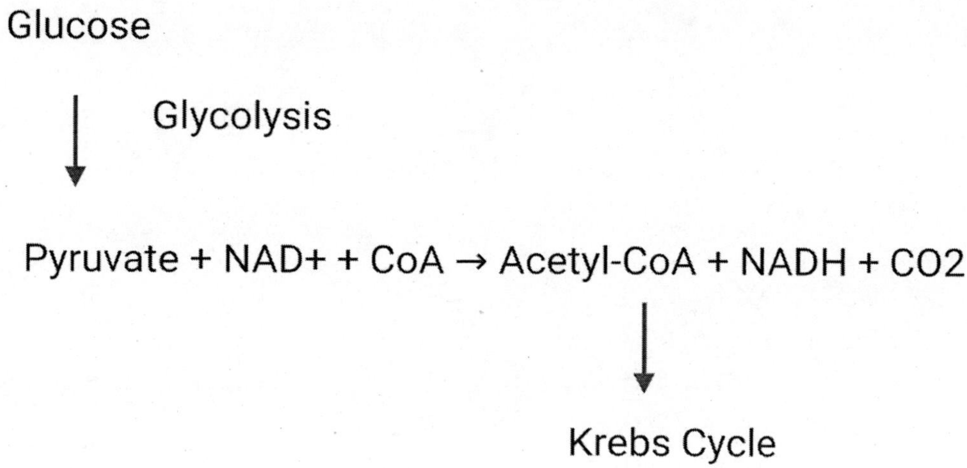

Fig. 8. 7 Connecting link between glycolysis and Krebs cycle

(Images created using BioRender® software)

The oxidation of pyruvate is the link between glycolysis and the Krebs cycle. In glycolysis, a 6-carbon glucose molecule is broken into two 2 molecules of pyruvates (3 carbons each). Due to the formation of 2 molecules, the link reaction occurs twice for each glucose molecule, producing two acetyl-CoA molecules, which can subsequently enter the Krebs cycle.

8.3.3 Krebs cycle/Citric acid cycle

The citric acid cycle refers to citrate, or its protonated form, citric acid, the first molecule formed during the cycle's processes. This sequence of reactions is also known as the tricarboxylic acid (TCA) cycle, due to 3 carboxyl groups on its first two intermediates, or the Krebs cycle, in honor of its discoverer, Hans Krebs.

It uses acetyl Co-A formed by the oxidation of glucose-derived pyruvate as its initial substrate and, through a sequence of redox reactions, extracts the majority of its bond energy as NADH, $FADH_2$, and ATP molecules. In addition, the reduced electron carriers NADH and $FADH_2$ produced by the TCA cycle transfer the electron to the Electron transport chain, and by oxidative phosphorylation in cellular respiration, ATP is generated.

Similar to the conversion of pyruvate to acetyl Co- A, the citric acid cycle occurs in the matrix of the mitochondria in eukaryotes. Whereas in prokaryotes, both of these processes take place in the cytoplasm. The citric acid cycle is a closed loop; the final step of the process restores the initial molecule. The cycle consists of 8 key phases, as shown in **Fig. 8.8**.

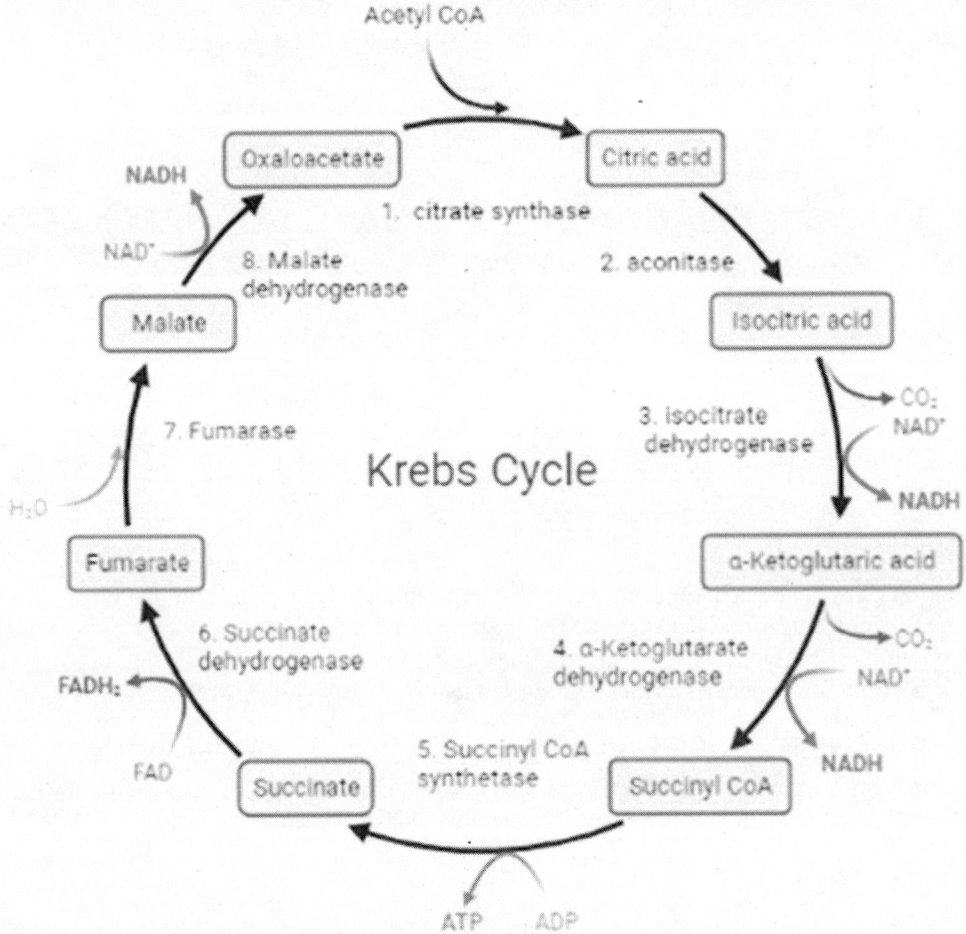

Fig. 8. 8 Reactions in Krebs cycle

(Images created using BioRender® software)

- In the initial step of the citric acid cycle, acetyl Co-A interacts with oxaloacetate, releasing the Co-A group and producing citrate, a six-carbon molecule.
- Citrate is transformed into its isomer, isocitrate, in the second step. This step involves the removal and subsequent addition of a water molecule.
- In the third step, isocitrate is oxidized, releasing a CO_2 molecule and leaving behind α-ketoglutarate, a five-carbon molecule. During this stage, NAD^+ is converted to NADH. Isocitrate dehydrogenase is the enzyme that catalyzes this step and is essential for regulating the rate of the citric acid cycle.

- The fourth step resembles the third step. In this instance, α-ketoglutarate is oxidized, converting NAD^+ to NADH, and a CO_2 molecule is released. The remaining four-carbon molecule combines with Coenzyme A to produce the unstable molecule succinyl CoA. α-ketoglutarate dehydrogenase, the enzyme that catalyzes this step, is also essential for regulating the citric acid cycle.
- In step five, the CoA of succinyl CoA is replaced with a phosphate group, which is then transferred to ADP to produce ATP. In some of the cells, GDP (guanosine diphosphate) is used in place of ADP to produce GTP (guanosine triphosphate). The resulting four-carbon compound is succinate.
- In step six, succinate is oxidized to generate fumarate, a four-carbon molecule. In this reaction, two hydrogen atoms and their electrons are transferred to FAD, forming $FADH_2$. This enzyme is buried within the inner membrane of the mitochondrion, where it allows the direct transfer of electrons from FADH2 to the electron transport chain.
- In step seven, water is added to the four-carbon fumarate molecule, transforming it into the four-carbon malate molecule.
- Malate is oxidized in the final phase of the citric acid cycle to oxaloacetate, leading to the regeneration of the initial four-carbon molecule. In this process, another molecule of NAD^+ is converted to NADH.

8.3.4 Electron transport chain

Like many other species, we require oxygen to survive. If you've ever tried to hold your breath for too long, you know that a lack of oxygen can lead to dizziness or even unconsciousness, and a prolonged shortage of O_2 can even result in death. But have you ever questioned why this is the case and what your body accomplishes with this oxygen?

So, how does oxygen play its role in this scenario? At the end of the electron transport chain, **oxygen** takes electrons and acquires protons to produce water. Chemiosmosis will cease to make ATP if there is insufficient oxygen to receive electrons (for example, if a person is not breathing in enough oxygen). Without sufficient ATP, cells cannot carry out the essential reactions for survival and may even die if oxygen is not supplied for a long enough period.

The electron transport chain is a collection of proteins and organic compounds in the mitochondrial inner membrane. In a sequence of redox reactions, electrons are continuously transferred from one member of the transport chain to another. The energy produced in these processes is collected as a proton gradient and used to produce ATP via a process known as chemiosmosis. The electron transport chain and chemiosmosis work together to form oxidative phosphorylation.

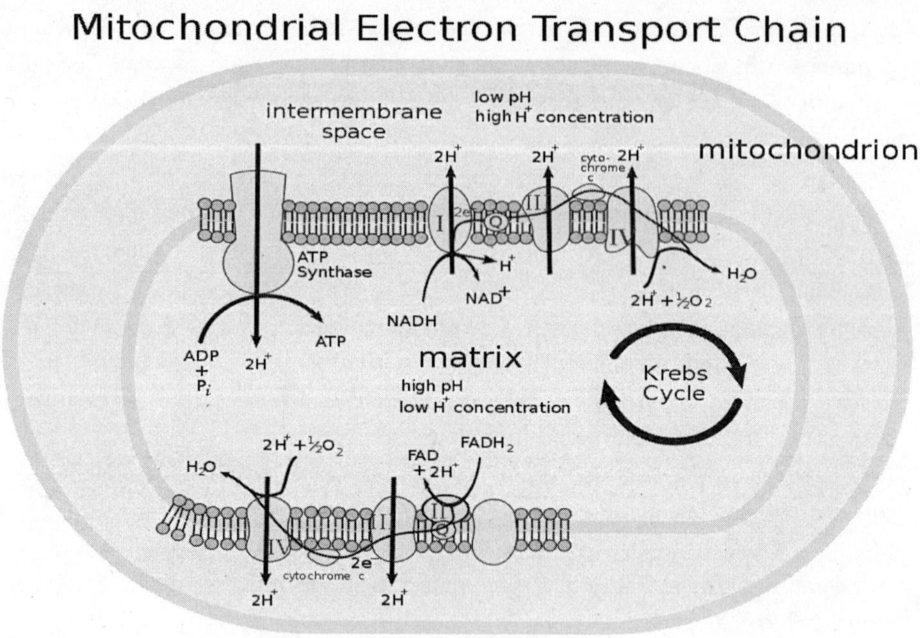

Fig. 8. 9 Electron transport chain

(Creative commons licenses)

Role of Electron transport chain

NADH and FADH$_2$ as electron carriers. Reduced electron carriers (NADH and FADH$_2$), produced in cellular respiration, transfer their electrons to molecules at the beginning of the transport chain. In the process, they turn back into NAD$^+$ and FAD, which can be reused for cellular respiration.

Transfer of electron and proton pumping. As electrons move along the chain, they move from a higher tô a lower energy level, releasing energy. Some energy is used to pump H$^+$ ions, moving them out of the matrix and into the intermembrane space. This pumping establishes an electrochemical gradient.

Gradient-driven ATP synthesis. As H$^+$ ions flow down their gradient and back into the matrix, they pass through an enzyme called ATP synthase, which harnesses the flow of protons to synthesize ATP.

Water formation by the splitting of oxygen. At the end of the electron transport chain, electrons are transferred to molecular oxygen, which splits in half and takes up H$^+$.

8.3.5 Photosynthesis

Food also provides the necessary precursor molecules for synthesizing new living matter and storing energy inside the chemical bonds of organic molecules. Some animals eat other creatures for sustenance, while others consume plants. In contrast, plants derive their energy directly from sunlight. Therefore, the sun is the ultimate source of the energy that animals acquire by eating plants or consuming creatures that have eaten plants.

Photosynthesis uses solar energy to generate Glucose from carbon dioxide and water. Oxygen gas is the byproduct of photosynthesis. Chlorophyll, carotenoids, and phycobilins are pigments used to absorb solar energy. This light energy is transformed into potential chemical energy during photosynthesis, and Glucose provides the metabolic energy for all cellular functions.

There are two types of photosynthesis, namely 1) **oxygenic** and 2) **anoxygenic**. Plants, cyanobacteria, and algae use oxygenic photosynthesis, while purple sulfur and green sulfur bacteria engage in anoxygenic photosynthesis. In the case of oxygenic water, oxygen serves as the electron donor, whereas nitrate, hydrogen sulfide, etc., are employed as electron donors in anoxygenic photosynthesis.

Oxygenic photosynthesis.

$$6CO_2 + 12\ HO_2 + \text{Light energy} \rightarrow 6C_6H_{12}O_6 + 6O_2 + 6HO_2$$

Anoxygenic photosynthesis.

$$CO_2 + H_2S + \text{Light energy} \rightarrow [CH_2O] + 2S + HO_2$$

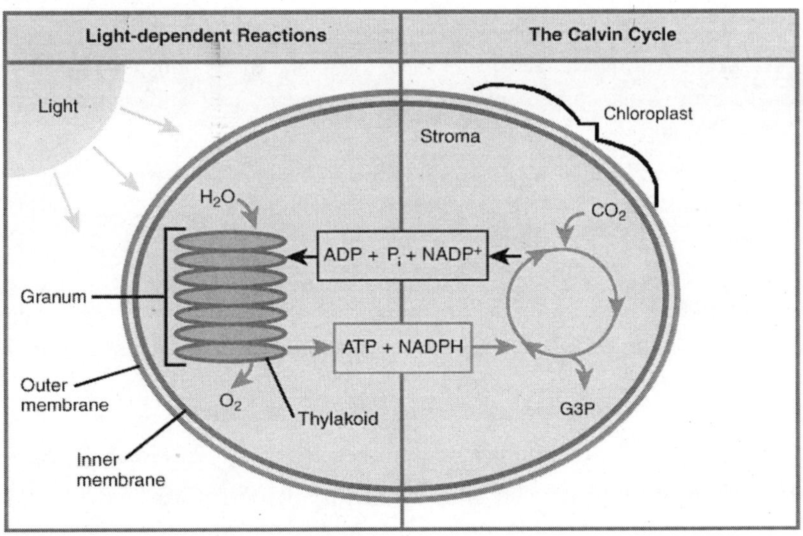

Fig. 8. 10 Overview of photosynthesis

(Creative commons licenses)

The light-dependent reactions in the thylakoid membrane require a constant source of light energy. This light energy is transformed into chemical energy by creating two compounds: ATP, an energy storage molecule, and NADPH, a reduced (electron-carrying) electron transporter. During this process, water molecules are also transformed into oxygen gas, the oxygen we breathe!

The Calvin cycle, commonly referred to as the light-independent process, occurs in the stroma and does not require light directly. Instead, the Calvin cycle uses ATP and NADPH from light-dependent processes to fix carbon dioxide and make molecules of glyceraldehyde-3-phosphate, or G3P, that combine to form Glucose.

Calvin Cycle or C2 cycle - CO_2 is fixed (incorporated into organic molecules) and is used to construct three-carbon sugars in the Calvin cycle. This process depends on ATP and NADPH, which are derived from photosynthesis. The light reactions occur in the thylakoid membrane, whereas light-independent reactions occur in the stroma (the inner space of chloroplasts).The reactions of the Calvin cycle can be categorized into three distinct phases: carbon fixation, reduction, and initial molecule regeneration.

Fixation of carbon - A CO_2 molecule joins with ribulose-1, 5-bisphosphate, a five-carbon molecule (RuBP). This process produces a six-carbon complex, which is very unstable and therefore breaks apart into 2 molecules of three-carbon molecules, 3-phosphoglyceric acid (3-PGA). This reaction is conducted by the enzyme rubisco, also known as RuBP carboxylase/oxygenase.

Reduction phase - 3-PGA is converted into a three-carbon sugar during the reduction phase of the Calvin cycle, which requires ATP and NADPH. This procedure involves two essential steps:

First, each molecule of 3-PGA obtains a phosphate group from ATP, transforming into 1, 3-bisphosphoglycerate, a molecule with two phosphate groups.

Second, the molecules of 1, 3-bisphosphoglycerate are reduced (gain electrons). Each molecule takes two electrons from NADPH and, in turn, loses one of its phosphate groups, transforming into glyceraldehyde 3-phosphate, a three-carbon sugar (G3P). This process generates the byproducts $NADP^+$ and Pi.

Regeneration - Some G3P molecules are utilized in the synthesis of Glucose, while others are recycled to regenerate the RuBP acceptor (see **Fig. 8.10**). Regeneration is dependent on ATP. It involves a complicated network of chemical processes.

Three turns of the Calvin cycle are required to produce a G3P molecule that can proceed to the production of Glucose. A G3P molecule contains three fixed carbon atoms; hence, two G3P molecules are required to construct a six-carbon glucose molecule. Therefore in total six rounds of calvin cycle, or 6 CO_2, 18 ATP, and 12 NADPH, are required to generate one molecule of Glucose.

Photosynthesis and cellular respiration are complementary processes in the living world.

Photosynthesis is performed by plants and photosynthetic microorganisms, which use the energy of sunlight to generate sugars and other organic compounds from the carbon atoms of atmospheric CO_2. These organic compounds serve as food for other organisms (animals and humans). During cellular respiration, these organisms use O_2 to oxidize food molecules, releasing the same carbon atoms in the form of CO_2 back into the atmosphere. In the process, the organisms obtain the chemical bond energy needed to survive, as described in **Fig. 8.11.**

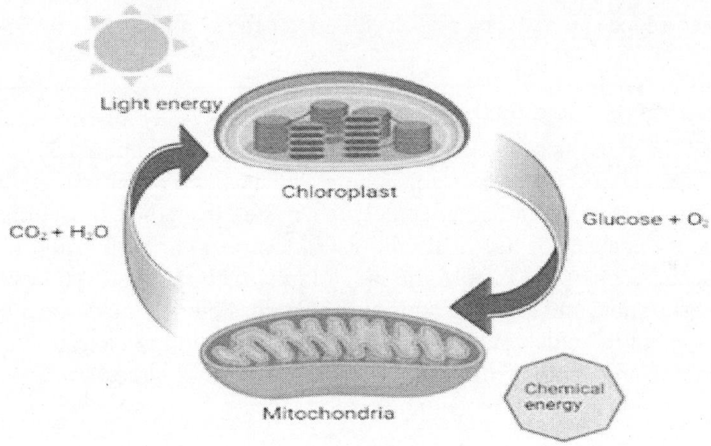

Fig. 8. 11 Relationship between photosynthesis and cellular respiration

8.3.6 Difference between cellular respiration and photosynthesis

It is essential to keep in mind that the processes of photosynthesis and respiration are carried out in the opposite order, despite being very similar. For instance, the inputs that go into the process of photosynthesis are water and carbon dioxide, and the by-products that come out of them are Glucose and oxygen. On the other hand, during the process of cellular respiration, oxygen and Glucose result in the formation of water molecules and the emission of carbon dioxide as a by-product.

Table 8. 4 Differences in photosynthesis and cellular respiration

Parameter	Endergonic reactions	Exergonic reactions
Function	›tures the energy and converts it into Glucose, and stores it.	eases energy by the breakdown of Glucose.
Iolecules are involved in the reaction.	$CO_2 + H_2O$ + Light	O_2 + Glucose.
Metabolic process	Anabolic	catabolic
›cation of reaction in the cell	chloroplast	mitochondria
Source of Energy	Sunlight	Glucose
Electron carrier	NADPH	NADH and $FADH_2$

Major occurrence	ts, photosynthetic bacteria, and algae.	Living organisms

The processes of cellular respiration and photosynthesis both play an essential role in maintaining the balance of the ecosystem's energy levels and are mutually beneficial. Light is used as a catalyst in the chemical reaction known as photosynthesis, in which carbon dioxide and water are combined to produce organic molecules. After then, the cell will use this organic substance as a source of sustenance for itself. In contrast, the process of respiration results in the creation of energy in the form of ATP from the breakdown of food. Photosynthesis is impossible without cellular respiration, and cellular respiration is impossible without photosynthesis. Neither one can happen without the other. Both photosynthesis and cellular respiration can be thought of as the "opposite" of one another in a number of ways. Photosynthesis is an anabolic process, while cellular respiration is a catabolic one.

UNIT SUMMARY

The branch of science that deals with the relationship between energy, heat, and work is called Thermodynamics. The main aim of the study of chemical thermodynamics is to learn the transformation of energy from one form into another and the utilization of various forms of energy. Metabolism is a set of all chemical reactions taking place in an organism. They involve a transformation of nutrients, in most cases catalyzed and regulated. Metabolic reactions are often coupled together to form metabolic pathways, where one substance is transformed, through a series of reactions, into another one. This process produces various intermediates, which can act as an initial substrate for another metabolic pathway.

EXERCISES

Multiple Choice Questions

1) Energy is stored in the form of________
   A. GTP
   B. ATP
   C. CTP
   D. TTP

2) ________ law of thermodynamics is used to understand the concept of energy conservation.
   A. First
   B. Second
   C. Zeroth
   D. Third

3) According to the second law of thermodynamics, the amount of disorder in the universe or an isolated system can only_________
   A. Decrease

B. Increase
C. Remain constant
D. Remain random

4) A downhill reaction is ______
A. Energetically favorable
B. Energetically unfavorable
C. Energetically sufficient
D. Energetically insufficient

5) Energetically beneficial reactions bring about the disorder by lowering the systems
A. Gibbs free energy
B. Enthalpy
C. Entropy
D. Free energy

6) ______ can create biological order by linking energetically unfavorable reactions with energetically favorable reactions.
A. Protein
B. Enzyme
C. DNA
D. RNA

7) The __________ is independent of concentration.
A. $\Delta G°$
B. $G°$
C. ΔH
D. $\Delta H°$

8) Favorable reactions are______
A. Exergonic
B. Endergonic
C. Both
D. None

9) Which of the best describes entropy
A. Useful heat energy needed to drive an engine
B. Heat that can be stored to be used at a later date
C. Random energy that escapes from the system
D. A state of order in a physical system

10) Which is an example of decreasing entropy in a closed system?
A. Boiling water
B. Freezing water
C. Cells in a body coming together
D. Ice melting

11) The process used in the conversion of pyruvate to acetyl CoA is
   A. Oxidative dehydration
   B. Decarboxylation
   C. Oxidative Phosphorylation
   D. Oxidative dephosphorylation
   E. None of the above

12) The release of energy obtained by the oxidation of Glucose is stored in the form of
   A. FAD
   B. ADP
   C. ATP
   D. NAD

13) A kinase is an enzyme that
   A. removes phosphate groups of substrate.
   B. uses ATP to add a phosphate group to the substrate.
   C. uses NADH to change the oxidation state of the substrate.
   D. removes water from a double bond

14) Anaerobic process after glycolysis is called
   A. Decarboxylation
   B. Kreb cycle
   C. Gluconeogenesis
   D. Fermentation

15) Krebs Cycle is termed the aerobic Phase of Respiration Because of
   A. Aerobic Conditions are essential for the continued operation of the Electron Transport System.
   B. It Consumes Oxygen.
   C. Oxygen acts as a Catalyst
   D. None of the Abov

Answers 1) B; 2) A; 3) B; 4) A; 5) D; 6) B; 7) A; 8) A; 9) C; 10) B; 11) B; 12) C; 13) B; 14) D; 15) A

Short Answer Type Questions

1. What are catabolism and anabolism?
2. Describe entropy
3. Thermodynamic first law
4. Thermodynamic second law
5. Explain Gibb's free energy
6. Explain Enthalpy
7. Change in free energy

8. What are endergonic and exergonic reactions?
9. How do cells obtain energy?
10. Explain pyruvate decarboxylation.
11. Elaborate photosynthesis.
12. Formation of lactic acid from pyruvate.

Long Answer Type Questions

1. Explain Cell creates order, then also follows thermodynamics law.
2. Explain that photosynthesis and cellular respiration are complementary in the living system.
3. Explain Gibb's free energy in brief.
4. Difference between cellular respiration and photosynthesis.
5. Balance sheet of glycolysis.
6. Explain cellular respiration.
7. Schematically explain glycolysis.
8. Schematically explain the Kreb cycle.
9. Explain the electron transport chain.
10. Write about the fate of pyruvate.

NUMERICAL PROBLEMS

1. Assume that an animal cell is a cube with a side length of 10 mm. The cell contains 10^9 ATP molecules that it uses up every minute. ATP is regenerated by oxidizing glucose molecules. After what amount of time will the cell have used up an amount of oxygen gas that is equal to its own volume? (recall that one mole contains **6×10^{23}** molecules. one mole of a gas has a volume of 22.4 liters.)

Ans. 30 ATP molecules are produced from each glucose molecule that is oxidized according to the reaction $C_6H_{12}O_6$ (Glucose) + $6O_2$ $6CO_2$ + $6H_2O$ + energy. Thus, one O_2 molecule is consumed for every five ATP molecules produced. The cell, therefore, consumes 2×10^8 O_2 molecules/min, which corresponds to the consumption of 3.3×10^{-16} moles (= $[2 \times 10^8] / [6 \times 10^{23}]$) or 7.4×10^{-15} liter (= $3.3 \times 10^{-16} \times 22.4$) each minute. The volume of the cell is 10^{-15} m^3 [= $(10^{-5})^3$], which is 10^{-12} liter. The cell, therefore, consumes about 0.7% (=$100 \times 7 \times 10^{-15}/10^{-12}$) of its volume of O_2 gas every minute or its own volume of O_2 gas in 2 hours and 15 minutes.

KNOW MORE

Each cell is estimated to generate and consume approximately 10,000,000 molecules of ATP per second.

Difference between NADH and NADPH: NADH (nicotinamide adenine dinucleotide) and the closely related molecule NADPH (nicotinamide adenine dinucleotide phosphate) carry energy in the form of two high-energy electrons plus a proton (H+), which together form a hydride ion (H–). When these activated carriers pass their energy (in the form of a hydride ion) to a donor molecule, they become oxidized to form NAD+ and NADP+. Differ in a single phosphate group. There is no effect on the electron-transfer properties it gives. NADPH is a slightly different shape from NADH to bind as substrates to different sets of enzymes and thereby deliver electrons (in the form of hydride ions) to different target molecules. NADPH operates chiefly with enzymes that catalyze anabolic reactions, supplying the high-energy electrons needed to synthesize energy-rich biological molecules. NADH, by contrast, has a special role as an intermediate in the catabolic system of reactions that generate ATP through the oxidation of food molecules. Inside the cell, the ratio of NAD+ to NADH is kept high, whereas the ratio of NADP+ to NADPH is kept low. This arrangement provides plenty of NAD+ to act as an oxidizing agent and plenty of NADPH to act as a reducing agent—as required for their special roles in catabolism and anabolism, respectively.

Firefly Flashes: Glowing Reports of ATP

Bioluminescence requires considerable amounts of energy. In the firefly, ATP is used in reactions that convert chemical energy into light energy. In the 1950s, from many thousands of fireflies collected by children in and around Baltimore, William McElroy and his colleagues at Johns Hopkins University isolated the principal biochemical components: luciferin, a complex carboxylic acid, and luciferase, an enzyme. The generation of a light flash requires the activation of luciferin by an enzymatic reaction involving pyrophosphate cleavage of ATP to form luciferyl adenylate. In the presence of molecular oxygen and luciferase, the luciferin undergoes a multistep oxidative decarboxylation to oxyluciferin. This process is accompanied by the emission of light. The color of the light flash differs with the firefly species and seems to be determined by differences in the structure of the luciferase. Luciferin is regenerated from oxyluciferin in a subsequent series of reactions.

REFERENCES AND SUGGESTED READINGS

1. Bruce Alberts, Dennis Bray, Karen Hopkin, Alexander Johnson, Julian Lewis, Martin Raff, Keith Roberts, and Peter Walter. Essential Cell Biology. Third Edition
2. Alberts Helmenstine, Anrie. (2020). Endergonic vs. Exergonic Reactions and Processes. Retrieved from https://www.thoughtco.com/endergonic-vs-exergonic-609258
3. Nelson, D. L.; and Cox, M. M.W.H. Freeman, and Company David L. Nelson and Michael M. Cox.Principles of Biochemistry.Lehninger. (2008). Principle of Biochemistry.5th edition.
4. Lodish, Harvey; Berk, Arnold; Matsudaira, Paul; Kaiser, Chri.(2003). Molecular Cell Biology. Fifth edition.

Dynamic QR Code for Further Reading

9 Microbiology

UNIT SPECIFICS

Through this unit we have discussed the following aspects:

- *Explained the difference between species and strain*
- *Identification and classification of microorganisms*
- *Ecological aspects of single-celled organisms*
- *Culturing techniques for organisms*

*The **practical applications of the topics are discussed** for generating further curiosity and creativity as well as improving problem solving capacity.*

*Besides giving a large number **of multiple choice questions as well as questions of short and long answer types** marked in two categories following lower and higher order of **Bloom's taxonomy**, assignments through a number of numerical problems, a list of references and suggested readings are given in the unit so that one can go through them for practice. It is important to note that for getting more information on various topics of interest some **QR codes** have been provided in different sections which can be scanned for relevant supportive knowledge.*

*After the related practical, based on the content, there is a **"Know More"** section. This section has been carefully designed so that the supplementary information provided in this part becomes beneficial for the users of the book. This section mainly highlights the initial activity, examples of some interesting facts, analogy, history of the development of the subject focusing the salient observations and finding, timelines starting from the development of the concerned topics up to the recent time, applications of the subject matter for our day-to-day real life or/and industrial applications on variety of aspects, case study related to environmental, sustainability, social and ethical issues whichever applicable, and finally inquisitiveness and curiosity topics of the unit.*

RATIONALE

The discipline will allow us to learn about every part of the organisms in order not only to understand their existence in the environments and also how they affect their surroundings and, in turn, other organisms nearby (human beings, animals, etc.).Microbiology is the most significant field in biology, making it possible to define how some microorganisms cause diseases, discover treatments for such diseases, and even use a few microbes for industrial applications.

PREREQUISITES

Biology class XI and XII

UNIT OUTCOMES

List of outcomes of this unit is as follows:

U9-O1: Explain the concept of single-celled organisms, species, and strains.

U9-O2: Discuss the classification of organisms and their identification through microscopy.

U9-O3: Explain the role of microbes in the environment

U9-O4: Explain the growth kinetics of bacterial culture.

U9-O5: Discuss the basis of microbial cultivation techniques along with sterilization methods

Unit-9 Outcomes	**EXPECTED MAPPING WITH COURSE OUTCOMES** (1- Weak Correlation; 2- Medium correlation; 3- Strong Correlation)				
	CO-1	**CO-2**	**CO-3**	**CO-4**	**CO-5**
U9-O1	3	2	2	1	1
U9-O2	3	3	1	1	2
U9-O3	3	2	2	1	1
U9-O4	3	2	1	1	-
U9-05	3	1	1	1	-

9.1 INTRODUCTION

Since the planet's earliest days, microorganisms have played a crucial role in shaping the planet's ecosystems and shaping the history of life on Earth. For obvious reasons, microorganisms are the first things that researchers look at on other planets when they investigate life. Ancient rocks and sediments contain fossils that indicate bacteria-like cells have been around for at least 3.5 billion years. This class of early microbes dominated life on Earth for the first two billion years. These ancient cells were not only small and straightforward, but they also lacked the specific intracellular structures required to perform their activities. These cells clearly did not have their DNA enclosed in a nucleus, often known as a karyon. The term prokaryotic*, which means "before the nucleus," is used to describe cells and microorganisms of this kind.

Microbiology is the branch of science that deals with the study of microbes. Microbes are microscopic living entities present everywhere. Most microorganisms are made up of a single cell and exhibit properties shared by all biological systems. These properties include reproduction, metabolic activity, growth, irritability, adaptability, mutation, and organization.

The history of microbiology begins with the contributions made by Louis Pasteur and Robert Koch in the latter half of the 19th century, which laid the groundwork for the field. Since then, many disease-causing bacteria have been discovered, and methods for minimizing their detrimental effects on humans have been devised. Additionally, methods have been identified for utilizing the activities of diverse microbes for the benefit of medicine, industry, and agriculture. For example, Antibiotics, such as penicillin, can be produced by fungi, namely molds.

An understanding of pathogenic organisms is essential for infection prevention. This chapter shows that this understanding is based on fundamental knowledge of microorganisms, including their identification, significance, and basic laboratory techniques. Bacteria, viruses, fungi, protozoa, algae, and helminths are the prominent microorganisms covered in this module.

9.2 CONCEPT OF SINGLE-CELL ORGANISMS

One can easily spot living things like plants, animals, or even mushrooms when one looks around. However, the environment consists of most living organisms which are invisible to the human eye. Cells are the living blocks of even the tiniest living thing. Some of these organisms are made up of only one cell and hence are called a single-celled organisms.

Unicellular creatures, the first forms of life on Earth, emerged from single cells millions of years ago. Digestion, excretion, and respiration are a few biological processes that occur inside a single cell in unicellular organisms. These are dubbed "microorganisms" because they are too small to be seen by the human eye—bacteria, protozoa, algae, fungi, etc.

Generally, prokaryotes are smaller than eukaryotes. In addition to lacking a well-defined nucleus, they also lack organelles, which are components in cells that are bound by one or more membranes. The mitochondria and Golgi complexes, among others, are organelles that perform respective functions like trafficking, food acquisition, producing energy, and protein synthesis. Prokaryotes execute the same duties as eukaryotes but lack the organelles necessary to carry them out.

9.2.1 Characteristics of Unicellular Organisms

Cells function differently in unicellular and multicellular organisms, although each cell in every organism contains numerous organelles or specialized cell components. These organelles are responsible for several cellular processes, as mentioned above. Included among unicellular creatures are bacteria, protists, and yeast. For example, Paramecium is a slipper-shaped, unicellular organism found in pond water. It absorbs nutrients from the water and digests them in organelles called food vacuoles. The cytoplasm transports nutrients from the food to the surrounding organelles, hence maintaining the health of the cell and the organism as a whole.

Types of Unicellular Organisms

There are two major types of cells: prokaryotic and eukaryotic. Bacteria and Archaea are single-celled creatures categorized as prokaryotes (pro = before; karyon– = nucleus). Animal cells, plant cells, fungi, and protists are eukaryotes (eu = true).

Prokaryotes - cell is a simple, unicellular organism with no nucleus or other membrane-bound organelles. This will soon become apparent to be drastically different in eukaryotes. Prokaryotic DNA is in the center of the cell, in a dark area known as the nucleoid. In contrast to Archaea and eukaryotes, bacteria have a cell wall constructed of peptidoglycan composed of carbohydrates and amino acids and a polysaccharide capsule. The cell wall provides an additional layer of protection, aids in maintaining the cell's shape, and prevents dehydration. The cell's ability to adhere to surfaces in its surroundings is enabled by its capsule. Some prokaryotes have flagella, pili, or fimbriae. Flagella are utilized for movement, while most pili are employed to exchange genetic material during conjugation, a sort of reproduction.

Eukaryotes - Comparable to how a large house is divided into numerous rooms with specific functions (bedrooms, baths, kitchen, living room, etc.), eukaryotic cells have numerous compartments with specialized functions, neatly separated by membrane layers. This structure allows each compartment to maintain its own conditions, those required to perform its function. Lysosomes, which operate as recycling centers for the cell, must, for example, maintain an acidic pH to dispose of cellular waste. Similarly, structures known as peroxisomes carry out oxidation reactions and produce hydrogen peroxide, both of which would harm the cell if not properly kept away in their own "chamber." Because eukaryotic cells can sustain several habitats within a single cell, they can perform complicated metabolic reactions that prokaryotes cannot. Indeed, it is a major reason why eukaryotic cells can grow many times larger than bacterial cells.

9.2.2 Characteristics of Multicellular Organisms

A multicellular organism is one made of several cells. Technically, the term "multi" refers to something that is more than one; therefore, "multicellular" denotes more than one. Multicellular organisms grow through cellular specialization and labor division. Cells become specialized in one process and are interdependent on one another to carry out other functions. For an organism to operate properly, all its cells must work together in harmony.

Cells of multicellular animals may also differ in appearance depending on the organelles required within the cell. Muscle cells, for example, have more mitochondria than most other cells, allowing them to manufacture energy for movement easily; pancreatic cells, on the other hand, must produce numerous proteins and have more ribosomes and rough endoplasmic reticula to meet this requirement. Although all cells include organelles, the amount and type of organelles present indicate how the cell works.

9.3 CONCEPT OF SPECIES AND STRAIN

9.3.1 Species

Species are the largest group of organisms in which any two individuals of appropriate sexes can produce fertile offspring through sexual reproduction. Species are also a fundamental unit of classification. Species can also be defined by their similar DNA sequence, morphology, karyotype, behavior, ecological niche, or sequential development pattern. The term "species" refers to a classification used in the field of biology to group together closely related organisms that exhibit similar traits and can procreate with one another. This idea of biological species is put to good use in the discipline of biology and several other academic subfields that are conceptually and methodologically related to biology. Nevertheless, there are over 20 additional distinct species concepts that need to be taken into consideration.

One example is the ecological species concept, which defines a species as a group of organisms framed by the resources on which they depend (in other words, their ecological niche). Another example is the genetic species concept, which considers all organisms capable of inheriting traits from one another within a common gene pool and the amount of genetic difference between populations of the same species. Both examples are examples of species concepts. The genetic species concept, much like the biological species concept, considers which individuals are capable of interbreeding and the amount of genetic difference between populations of the same species. However, the genetic species concept can also be used to estimate when the species first appeared on the earth.

9.3.2 Strain

- A strain is a subtype or genetic variant of a biological species.
- Strains in microorganisms typically originate from a single-cell colony of microorganisms.
- Strains are commonly found in virology, botany, insects, and experimental rodents and are often considered an inherently artificial concept because it describes a specific intent, such as genetic isolation.

9.4 IDENTIFICATION AND CLASSIFICATION OF MICROORGANISMS

The classification of organisms is referred to as taxonomy. Taxa are the groups that comprise the classification. It also includes classifying new organisms and reclassifying existing ones. Microorganisms are typically identified using binomial nomenclature, which consists of two words that refer to the genus and the species. Microorganisms' names are written in Latin. The first letter of the genus name is always capitalized. A domain is the first, largest, and most inclusive group into which organisms are classified, and it is divided into three subgroups: bacteria, archaea, and eukarya. This first classification determines whether an organism is a prokaryote or a eukaryote. A kingdom is a name given to the second largest group. Prokaryota (e.g., archaea and bacteria), Protoctista (e.g., protozoa and algae), fungi, Plantae, and Animalia, are the five major kingdoms described. Bacterial classification uses the taxonomic ranks kingdom, division, class, order, family, genus, and species. Bacteria, archaea, fungi, protozoa, algae, and viruses are the six major groups into which they can be classified.

Bacteria -A single-celled creatures. Because the cells are nucleon-free, they are referred to as prokaryotic. They come in four different shapes: the rod-shaped bacillus, the spherical coccus, the spiral-shaped spirilla, and the vibrio (curved shape). Most bacteria divide via binary fission, contain peptidoglycan cell walls, and may have flagella for mobility. One of the key characteristics used to categorize these creatures is the variation in the structure of their cell walls. **When using Gram staining, bacteria can be classified as Gram-positive or Gram-negative based on how their cell wall structure stains.** Gram positive bacteria have thick peptidoglycan layer while peptidoglycan layer is thin in gram negative bacteria. The fundamental idea is based on the ability of bacterial cell wall to retain the primary stain. Bacterial cells possessing thick peptidoglycan layer resist the decolorization of primary stain and hence appear violet or purple. On the other hand, bacterial cells with thin peptidoglycan layer have less cross-linking and lose the primary stain during decolorization, therefore appear pink or red because of the counterstain. In Gram staining, the initial step involves staining with primary stain i.e. crystal violet dye. In the next stage, also known as fixing the dye, Gram's iodine is used as a mordant which forms a crystal violet-iodine complex to prevent the dye from being easily removed. A decolorizer, often a mixture of ethanol and acetone, is then used to remove the dye. After decolorization, a counterstain safranin is used to stain the decolorized cells pink. Bacteria are further classified into three types based on their response to gaseous oxygen: aerobic (living in the presence of oxygen), anaerobic (living without oxygen), and facultative anaerobes (can live in both environments).

Fungi - Eukaryotic cells include mushrooms, molds, and yeasts, which are fungi (with a true nucleus). Most fungi have many cells, and chitin makes up their cell walls. They absorb organic material from their surroundings (decomposers), form symbiotic partnerships with plants (symbionts), or engage in destructive interactions with hosts to gain nutrition (parasites). They produce hyphae, distinctive filamentous tubes that aid in material absorption. The group of hyphae is referred to as mycelium. Fungi release spores to proliferate.

Protozoa - Aerobic unicellular eukaryotes include protozoa. They have a nucleus, intricate organelles, and specific features that allow them to absorb or consume nutrients. They are the most numerous, biomass-producing, and diverse group of living things on the planet. Cellulose makes up the walls of their cells. Traditional protozoa classifications are based on how they move: ciliates have tiny hairs that beat to produce movement, amoeboids have false feet or pseudopodia used for feeding and locomotion, and sporozoans are non-motile. Flagellates produce their own food and propel themselves forward using their whip-like structure.

Algae - Algae are photosynthetic organisms belonging to Kingdom Protista. The size of algae can range from extremely small unicellular algae such as chlorella to enormous kelps that can grow to a length of up to 200 feet. Photosynthesis is how algae obtain their nourishment. They produce oxygen and carbohydrates that other species need while residing in water, moist soil, and rocks. Green land plants are said to have originated from cyanobacteria. Chlorophyceae, also known as green algae; Phaeophyceae, often known as brown algae; and Rhodophyceae, referred to as red algae, are the three primary types of algae classified based on pigments present, the chemical nature of assimilatory products, and other factors. Reproduction in algae can take place either sexually or asexually, using a variety of different reproductive methods. Algae reproduce in both sexual and asexual ways involving various reproductive strategies. In minor algal species, asexual reproduction proceeds through regular cell division or fragmentation, without the union of cells or the fusion of separate genetic materials. Superior algae reproduce through the release of spores. Meiosis is the process of sexual reproduction that uses genetic material from two separate parent cells. Different environmental occurrences have an impact on and control sexual reproduction.

Viruses - A protein coat surrounds the nucleic acid core (DNA or RNA) of viruses, which are noncellular organisms. Viruses are regarded to be microbes even though they are not considered to be living things. Outside of a host cell, viruses cannot procreate and cannot metabolize on their own. Disease-causing viruses frequently infect prokaryotic and eukaryotic cells. Like cellular organisms, based on phylogenetic characteristics, viruses can be classified according to morphology, nucleic acid type, mode of reproduction, host organisms, type of disease they cause. To classify viruses, two main classification systems are used 1. International Committee on Taxonomy of Viruses (ICTV), and 2. The Baltimore Classification system. According to ICTV, 6 orders have been established to date – Caudovirales, Herpesvirales, Mononegavirales, Nidovirales, Picornavirales, and Tymovirales. The Classification of Viruses as per the Baltimore Classification system is based on a combination of the following characters – their nucleic acid (DNA or RNA), strandness (single strand or double strand), sense (positive sense or negative sense), and method of replication.

Multicellular Animal Parasites - The flatworms and roundworms together, referred to as the helminths, are a class of eukaryotic creatures. Since they are large enough to be visible to the naked eye, they are not microorganisms by definition, yet they spend a portion of their life cycle as microscopic organisms. Due to their therapeutic significance, parasitic helminths are frequently discussed with other types of microorganisms.

9.4.1 Identification of Microorganisms

The correct identification of microorganisms is essential for microbiologists and scientists involved in research and industries. The identification of microorganisms is primarily dependent on phenotypic characteristics. However, advanced genotypic methods such as fingerprint and sequence-based techniques have recently been used. In this section, we will learn about primary methods for identification.

Microscopy: Morphology and staining reactions of organisms are used to classify unknown species. Gram staining distinguishes gram-positive and negative microorganisms as shown in **Fig. 9.1.**

Cultural characteristics: The appearance of colonial growth on nutrient agar is distinctive. Diameter, outline, elevation, translucency (clear, translucent, opaque), and color are noted.

Biochemical characteristics: Species that lack morphological and cultural distinctions may have metabolic differences. When glucose, lactose, sucrose, mannitol, etc., are the sole carbon supply, the organism's ability to create acidic and gaseous end-products is tested.

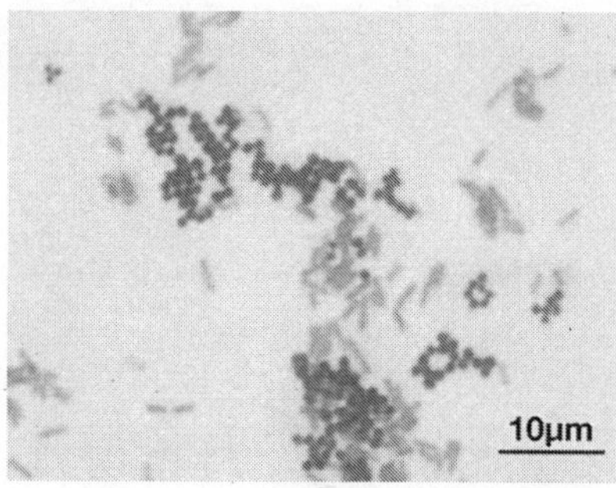

Fig. 9. 1 Gram-positive (purple) and Gram-negative (pink)

Source: Wikimedia commons (Creative common license)

9.5 MICROSCOPY

The microscope helps to observe microorganisms' traits. Bacteria, cells, and more are studied through **microscopes**. This technology allows biologists to investigate cells and living organisms. Microscope sizes and uses vary. Light and electron microscopes are popular. Each microscope has unique features and uses. Both light and electron microscopes employ radiation to create images the human eye can't.

Microscopes work by magnifying the microscopic workings of the world by magnifying a small-scale field of view. Magnification is just to enlarge the size of the object. The magnification depends on the magnifying power of the lenses used in the microscope. In microscopy, "resolution" refers to a microscope's ability to identify fine details. This is the minimal distance between two points of a specimen that may be distinguished as different entities. Either by the microscope camera or by the observer. Refer **Table 9.1** for the difference between light and compound microscope (**Fig. 9.2** and **9.3)**

Table 9. 1 Difference between Light and Electron Microscope

Light Microscope	Electron Microscope
Uses light (approx 400-700 nm) as an illuminating source	It uses electron beams (approx 1 nm) as an illuminating source
Lower magnification and resolution than an electron microscope	Higher magnification and resolution
Specimen preparation takes about a few minutes or an hour	Specimen preparation takes several days
Both live and dead specimens can be seen	Only dead and the dried specimen can be seen
The image formation depends upon the light absorption from the different zones of the specimen.	The image formation depends upon the electron scattering.
Inexpensive and requires a low maintenance cost	Expensive and high maintenance

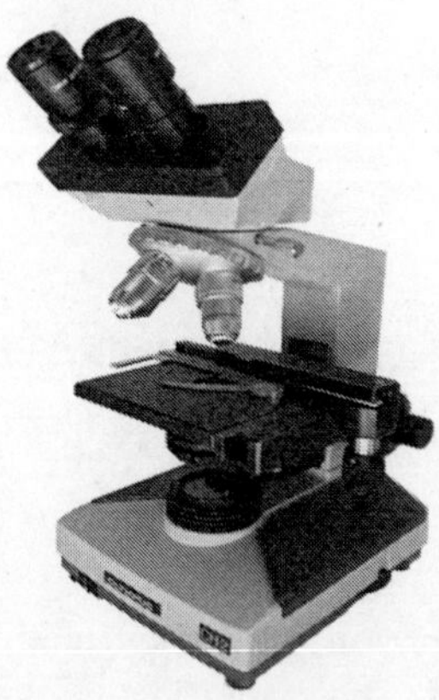

Fig. 9. 2 Compound Microscope

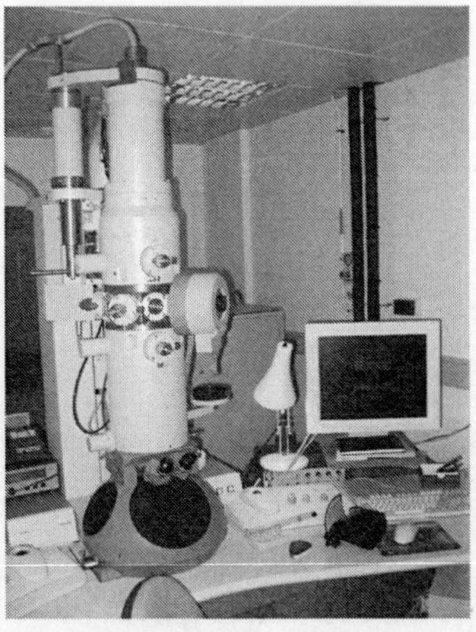

Fig. 9. 3 Electron microscope

Source: Wikimedia commons (Creative common license)

9.6 ECOLOGICAL ASPECTS OF SINGLE-CELLED ORGANISMS

Microbial ecology deals with the relationship of microorganisms with one another and their environment. Microbes are found all around the planet, including extreme conditions like acidic lakes, hydrothermal vents, etc. The biotic and abiotic factors both contribute to the development of the microbial community. Microbial processes occur in soil, air, or water. Symbioses, biogeochemical cycles, and the interaction of microorganisms with anthropogenic factors like pollution and climate change are all included in this field of study. Ecology studies the interaction of organisms in an area with the surrounding environment. This interaction constitutes an overall adaptation of the organisms to their environment, including the continuity of species. These interactions can be inter-specific (interactions with different species) or intra-specific (interactions between the same species). There are five types of interactions between other species, as listed below:

- Competition and Predation
- Commensalism
- Parasitism
- Mutualism
- Amensalism

Competition and Predation - When one entity hunts another animal to suffice its nutritional requirements, it is referred to as predation. A predator is an entity that hunts its prey. For example, a snake eats a frog. Here snake is the predator, and the frog is its prey. Competition, on the other hand, is when populations or even an individual compete for food resources. It is often referred to as exploitative or consumptive competition. For example, competition for territory.

Commensalism - It is an imbalanced type of interaction wherein one entity benefits while the other is neither harmed nor benefited. There are four types of commensal associations.

Inquilinism – An entity occupies the living habitat of another species (burrow, nest)

Chemical commensalism – A bacteria produces a chemical that nurtures another bacteria

Phoresy – An organism tentatively attaches itself to another entity for transportation requirements.

Metabiosis – One entity is dependent on the other for survival

Parasitism - One entity benefits from other entities and is harmed but not necessarily killed. The entity that is harmed is the host, and the one benefit is the parasite. When the host is killed, this type of behavior is referred to as parasitoidism. These parasites can live on the surface of the host, often addressed as ectoparasites (fleas, leeches), while endoparasites live inside the host. Endoparasites can be subdivided into intracellular parasites (live inside cells) and intercellular parasites (live in spaces between cells).

Mutualism - Both species involved in the interaction are benefited. These interactions take place in three patterns:

Facultative mutualism – Species survive on their own under favorable conditions

Obligate mutualism – One species is dependent for survival on the other

Diffusive mutualism – One entity can live with multiple partners

Amensalism - In this type of interaction, when one population finds itself in danger, the other population is not majorly affected. For instance, Tall and wide plants hinder the growth of comparatively smaller plants. Some plants even secrete substances that repress the growth of nearby plants to remove competition.

9.6.1 Role of microorganisms in the ecosystem

1. Produce airborne oxygen
2. Convert nutrients contained in organic materials into inorganic forms
3. Convert atmospheric nitrogen into useful forms.
4. Some microorganisms help in efficient absorption of nutrients from the soil by associating with the root hairs.
5. Allow herbivore animals to get nutrition from poor-quality food

9.7 CULTURING OF MICROORGANISMS IN THE LAB

In a laboratory, microorganisms are cultivated in a growth environment containing the required nutrients called a medium. The medium's composition could be entirely chemical (i.e., a chemically defined media), organically based, or living things like fertilized eggs. Microorganisms growing in or on such a medium form a culture.

If more than one type of organism is present, it is referred to as a mixed culture. If only one type of organism is present, it is referred to as a pure culture. The culture medium must be sterile when first used, which means it must be devoid of any life when the bacterium is introduced. To work with microorganisms in the laboratory, it is desirable to obtain them in pure cultures. Microbial cultures can be preserved at low temperatures as it reduces the rate of metabolism. This can be achieved by two methods deep-freezing and freeze-drying. To deep freeze microorganisms, they are submerged in a liquid and swiftly frozen at -50°C or lower. In an instrument used for freeze-drying (lyophilization), water is removed by vacuum after the microbial suspension has been frozen. The microorganisms can be kept alive for a very long time in this culture, which has the consistency of powder.

The isolation and inoculation techniques are used prior to adding the microbial strain to the culture media.

9.7.1 Inoculation

In microbiology labs, mounting microbial cultures on a culture media is a fundamental procedure. A platinum or nichrome wire inoculation loop with a loop at one end is used for the procedure. It is primarily utilized in culture plate and streaking procedures. The term "inoculum" refers to the tiny sample transmitted from the culture.

9.7.2 Isolation

A particular microbial strain can be isolated using the microbiological technique of isolation by cultivating the microbes in a specific culture medium. The process must be performed numerous times to completely purify the microbial strain and generate a pure culture. After that, the microbial strain can be seen in culture plates as distinct/isolated individual colonies, as shown in Fig. 9.4.

The isolation of the organisms (mainly bacteria) into a pure culture from a mixed population is accomplished using the streak plate technique. As the original sample is spread out over consecutive quadrants, fewer organisms are found there. The number of organisms reduces when the original sample is diluted by streaking it over subsequent quadrants. By the third or fourth quadrant, just a few organisms are typically observed, resulting in discrete colony-forming units (CFUs). When these lone bacterial cells divide and generate tens of thousands of other bacteria, they form an isolated colony. Well-isolated colonies can be selected and re-streaked on new agar plates to produce pure cultures. This method helps separate organisms in a mixed culture or when we need to study the colony morphology of an organism.

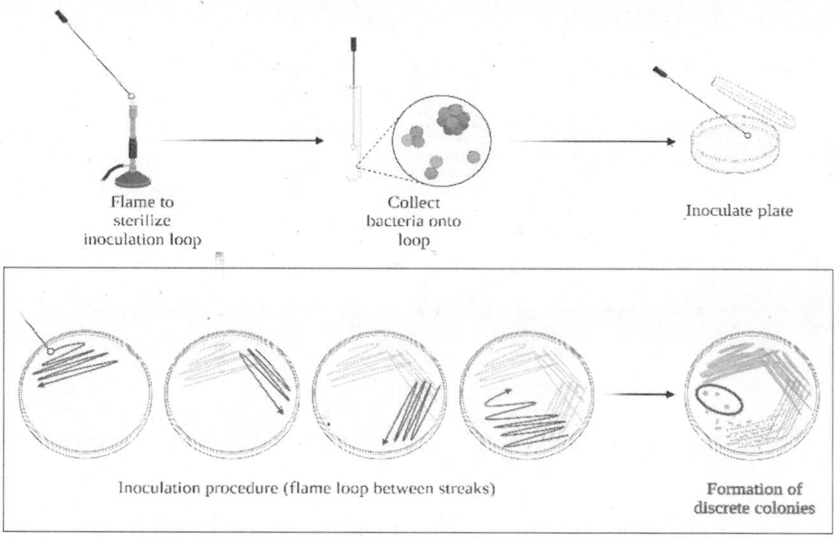

Fig. 9. 4 Streak Plate Technique

(Images created using BioRender® software)

9.7.3 Spread plate method

In this procedure, a very little amount of the liquid microbe suspension is put onto the media-containing plate's hardened surface. An L-shaped glass rod is then used to spread the liquid uniformly throughout the plate's surface, as shown in **Fig. 9.5**. This is done to collect individual colonies of microorganisms and count the quantity of microbes in the population. By counting the colonies on the plate and multiplying that total by the dilution factor, one can calculate the total number of viable organisms that were initially present in the sample.

CFU/ml = (number of colonies x dilution factor) / volume of culture plated.

The spread plate method can be used to count the number of bacteria present in a sample. The spread plate technique is most frequently utilized to isolate and identify a variety of microbial flora present in environmental samples like soil, food, or any other materials.

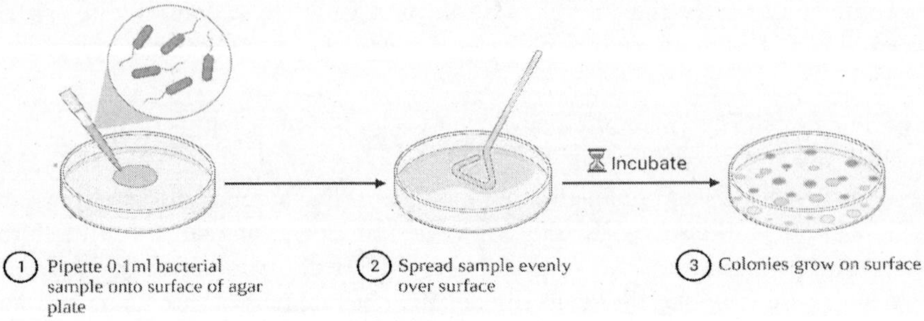

Fig. 9. 5 Spread Plate Technique

(Images created using BioRender® software)

9.7.4 Pour plate method

This technique involves pipetting a serially diluted microbe suspension onto a sterile Petri plate. After which, the culture media is liquefied and added to the dish. After the media has been set, the culture plate is incubated to check for specific bacterial growth, as shown in **Fig. 9.6**. It is done to gauge the number of live bacteria in a microbiological suspension. Both the top and the interior of the medium will exhibit growth. The following formula is used to calculate the number of microorganisms in a given test sample:

CFU/mL= CFU * dilution factor * 1/aliquot

The ideal count for accurate counting should fall between 30 and 300 colonies per plate.

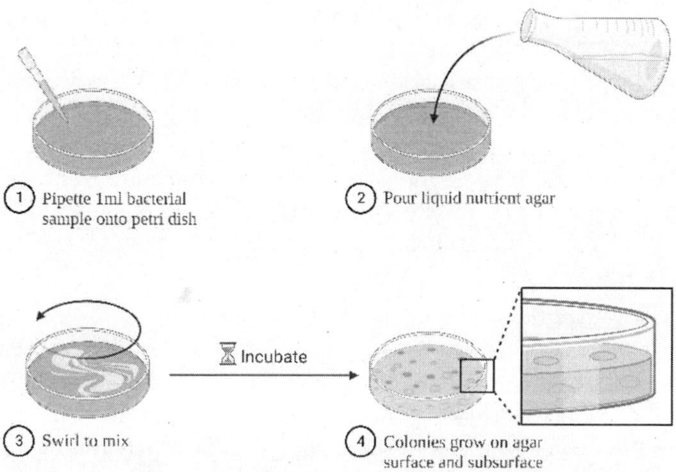

Fig. 9. 6 Pour Plate Technique

(Images created using BioRender® software)

9.8 STERILIZATION AND MEDIA COMPOSITIONS

9.8.1 Sterilization

Microorganisms play an important role in human welfare. Along with useful microorganisms, there are also pathogenic organisms detrimental to human health. Therefore, it is necessary to research microbial strategies to exploit the microbial population effectively. The potential of these can be brought to use only when they are studied in their pure form. Louis Pasteur was the first person to develop the process of sterilization.

Sterilization is known as killing all microorganisms (bacterial, viral, and fungal) with either physical or chemical agents. Sterilization is used to prepare culture media, reagents, and equipment where a sterile condition is required. There are three primary methods for killing, removing, or inhibiting organisms. They are 1) to destroy pathogens and prevent infection, 2) to prevent spoilage of food and other commodities, and 3) to avoid contaminating materials used in laboratory pre-culture operations, as well as to avoid interfering with numerous industrial processes that rely on pure cultures.

The following methods are used to sterilize microbiology laboratories.

Physical methods, such as the use of heat, filters, and radiation

A chemical process, i.e., using chemicals Sterilization by heat

Dry heat sterilization - Sterilization of inoculation loops or needles is accomplished by heating to 'red' in a Bunsen burner or spirit lamp flame. Dry heat sterilization is one of the most efficient and preferred methods of sterilization, employing hot air to kill or deactivate all forms of life within the chamber of an industrial oven. The heat that is transferred to the item denatures the proteins of all bacterial spores, fungi, viruses, prions, and biological agents in general. In a hot air oven, sterilization is performed at 160°C and held for one hour. At this temperature, spores are killed, and this is the most common method of sterilizing glassware, swab sticks, pestle and mortar, mineral oil, and other items. It is essential to remove any moisture from the air blown into the oven chamber, as moisture can interfere with the denaturation of proteins. *Bacillus atrophaeus* spores should be used to monitor dry heat sterilization because they are more resistant to dry heat than *Geobacillus stearothermophilus* spores.

Wet heat or moist heat sterilization - Boiling at 100°C for 30 minutes in a water bath achieves moist heat sterilization. This method can sterilize syringes, rubber goods, and surgical instruments. This method kills almost all bacteria and certain spores. Steaming at 100°C for 20 to 30 minutes at normal atmospheric pressure is more effective than dry heat at the same temperature because bacteria are more susceptible to moist heat. Steam has greater penetrating and sterilizing power because more heat is given up during condensation. One of the most common sterilization methods is autoclaving. In autoclaving, steam at temperatures above 100°C is used. This is accomplished by applying more pressure. The autoclave is closed and made airtight for pressure development, and 121°C temperatures are reached at 15 lbs per square inch pressure, **Fig. 9.7**. This temperature is given as sterilizing holding time for another 15 minutes. Pasteurization is another moist heat sterilization method that works below 100°C heat. This method is used to heat milk and other liquid foods. The product is kept at a specific temperature and time to kill any pathogenic bacteria that may be present. This process does not eliminate the entire organism, including spores.

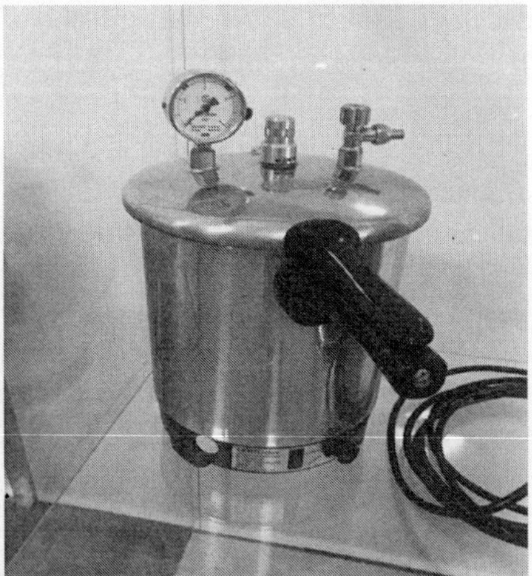

Fig. 9. 7 Autoclave

Source: Wikimedia commons (Creative common license)

Filtration - This sterilization method is used for media that are particularly heat-labile in nature. For example, serum in culture media gets easily coagulated by heat. If the study requires bacteria-free filtrates, 0.45 micron-sized filter membranes are used; if the study requires a viral particle-free solution, 0.22 micron-sized filter membranes are used. The principle of filter sterilization is based on membrane pore size and bacterial cell size. If the contaminants are smaller than the intended particle size, reduce the pore size of the membrane and capture the product while the impurities pass through the membrane. Filters can be added in parallel or series to increase system versatility. Throughput increases when a filter with the same pore size is added in parallel. Separation of various microorganisms is achievable if a filter with varying pore sizes is introduced in series. It is a recommended method of sterilizing heat-sensitive liquids and gases without exposing them to denaturing heat is filtration. It simply eliminates contaminated bacteria rather than destroying them. It is the preferred approach for sterilizing heat-sensitive antibiotic solutions, hazardous compounds, radioisotopes, vaccines, etc.

Ultraviolet Radiation - Not all types of UV light are effective for sterilizing. UV is classified into three kinds based on decreasing wavelengths and increasing energy. UVA, UVB, and UVC are the three types. Only UVC (100-280nm) provides enough energy to successfully destroy germs in UV sterilization. UVC at 254 nm has been demonstrated in studies to be efficient against all foodborne pathogens, natural microbiota, molds, and yeasts. It prevents the replication and transcription of DNA by causing thymine-thymine dimerization in the DNA. Because microorganisms vary in size and shape, which affects their UV absorption, the time required to kill each species varies. This method is used to sterilize the surface of laminar airflow, **Fig. 9.8,** biosafety cabinets, and in some cases, laboratories. UV light is ineffective in penetrating glass, dirt films, water, and other substances.

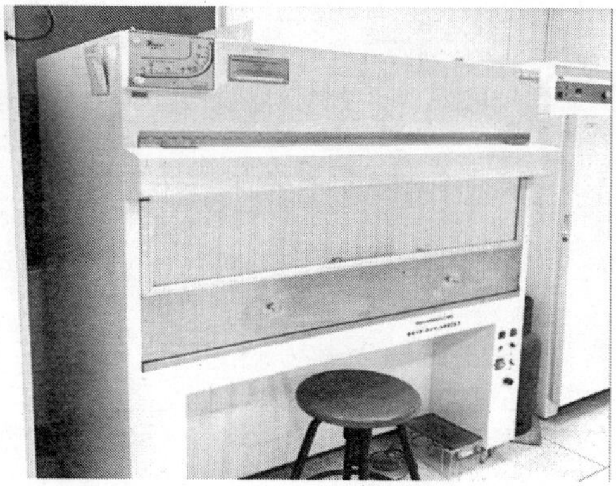

Fig. 9. 8 Laminar Air Flow Source

Wikimedia commons (Creative common license)

9.8.2 Types of media

When living in their natural habitat, microorganisms get the nutrients they need for their metabolic processes from the outside environment. In contrast, an artificial environment can be created in laboratories by supplying the necessary nutrients to the organisms being studied in the form of growth media. The microorganisms' nutritional needs can range from a relatively small number of simple inorganic molecules to a large number of complex inorganic compounds. This incredible range of possibilities is reflected in the kinds of media that can be produced. When it comes to cultivating microorganisms and determining their identities, a wide variety of media are utilized. The amount of nutrients that are present in media and the consistency of that media can vary greatly, and each type of media can be specifically created for a different purpose.

General media: Nutrient broth, a solution containing proteins, salts, and growth stimulants, is a common microbiological medium. Agar solidifies a medium but adds no nutrition. Liquid, semisolid, and solid media are generally used, as shown in **Fig. 9.9**. Solid and semi-solid media contain agar, whereas liquid media don't. In fermentation research and biochemical tests, liquid media can proliferate several microorganisms. Semi-solid media can be utilized for fermentation, bacterial motility, and anaerobic growth. Solid media are used to observe microorganism surface development and isolate pure cultures. Depending on the needs of specific bacteria, many types of culture media have been produced.

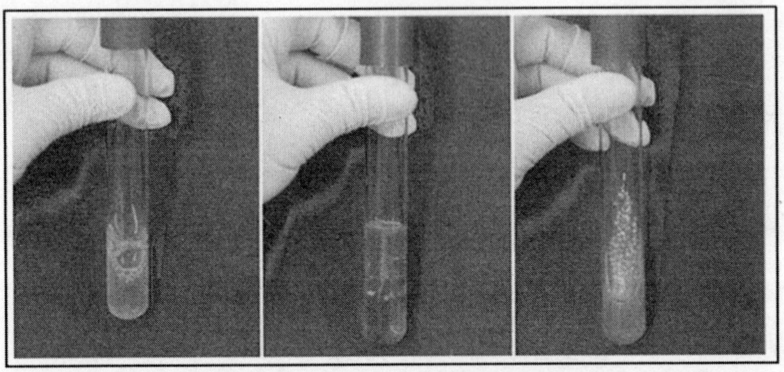

Fig. 9. 9 Bacterial growth in liquid, solid and semi-solid media

(Source: Coursehero)

Minimal and supplementary media: Minimal media contain minimal nutrients for wild-type organism development. For pure culture isolation, the minimal medium contains glucose, inorganic salts, and water. Supplemental media are minimal media that contain an amino acid or sugar for culture auxotrophs.

Synthetic and complex media: Synthetic media have known chemical compositions. Examples – are nutrient broth, tryptic soy broth, and MacConkey agar.

Enriched media: Enriched media contains a component that only certain organisms may use to flourish. Some have selected features. Blood agar is enriched media because it grows fastidious microorganisms.

Selective media: A selective medium is created by adding specific substances to a culture medium that allow one group of bacteria to grow while inhibiting the growth of others. Mannitol agar is a popular selective agar.

Differential Media: Differential or indicator media identify one microorganism kind from another. In these media, the biochemical features of a microbe growing in the presence of nutrients or indicators—such as neutral red, phenol red, eosin Y, or methylene blue—are used to visually identify a microorganism's traits. Molecular scientists use this media to identify recombinant bacterial strains and microorganisms. Examples of differential media include blood agar, MacConkey, etc.

Transport media: It's a buffer solution comprising peptone, carbohydrates, and other nutrients (excluding growth factors) to sustain bacterial viability throughout transport. For gonococci, use Stuart medium.

Anaerobic media: It supports anaerobic bacteria development. Robertson's cooked meat media. (Refer **Fig. 9.10** for bacterial growth on different growth media)

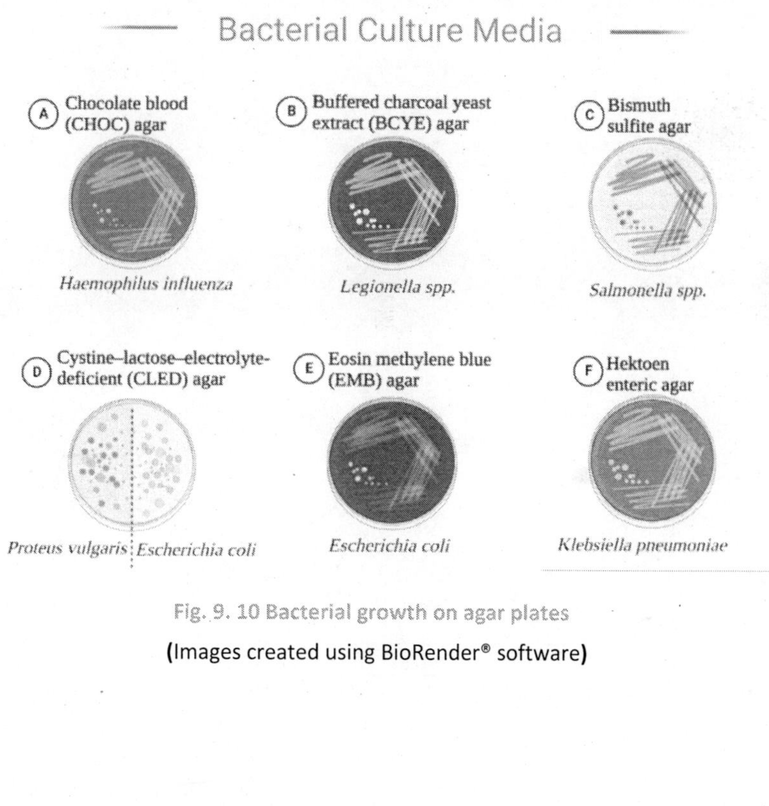

Fig. 9. 10 Bacterial growth on agar plates

(Images created using BioRender® software)

9.9 GROWTH KINETICS

Bacterial growth is associated with many complex anabolic and catabolic processes. An organism carries out these processes when allowed to grow under nutrient-rich conditions. When microbes receive nutrients and the necessary environmental conditions, they become metabolically active and proliferate. Development occurs on two levels. On one level, a cell synthesizes new cell components and grows in size, while on another level, the population of cells rises. Throughout their development cycles, bacteria reproduce several times resulting in a substantial increase in population size. Bacteria reproduce asexually through a mechanism known as binary fission. During this process, chromosomal DNA replicates, and then the bacterial membrane and cell wall grow inward to meet and divide the cell. The two cells separate, concluding the process. This capacity for multiplication, which increases population size through cell division, is crucial for microbial control, infectious disease, and biotechnology. Colony counting or UV-visible spectroscopy techniques can measure bacterial cells. Cell count versus incubation time defines the growth curve, **Fig. 9.11**. In the following sections, we will focus mostly on the growth features of bacteria typical of single-celled microorganisms.

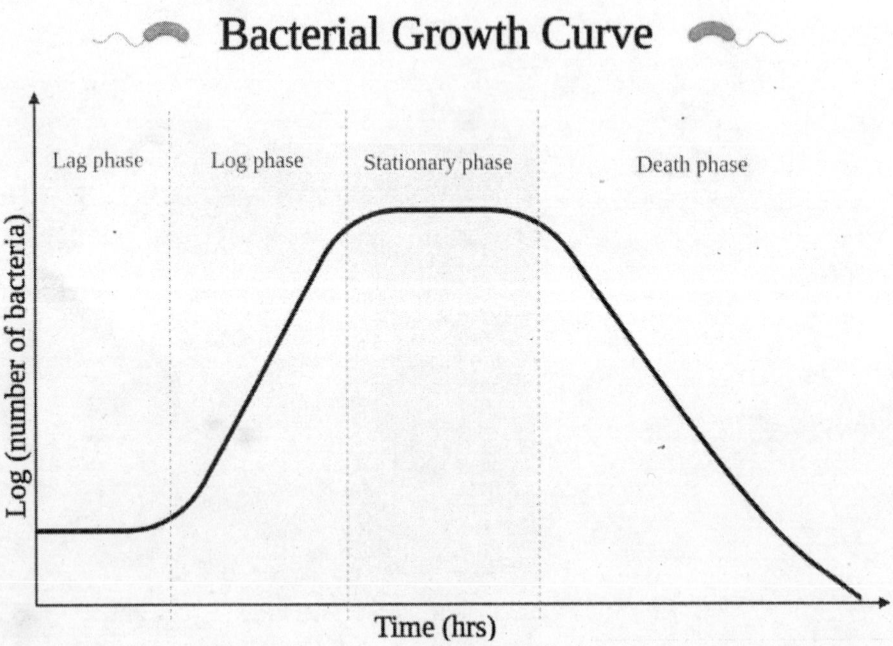

Fig. 9. 11 Growth Curve

(Images created using BioRender® software)

This section examines the four unique stages of the resulting curve.

Lag phase: When microorganisms are introduced into fresh growth media, they do not directly start to reproduce. It is a phase where cells are adjusting to the new growth conditions, and so also called the adaptation period. In the lag phase, bacteria are not metabolically inactive but repair themselves, produce enzymes, and synthesize chemicals required for carrying out physiological and biochemical processes. It is a phase where bacteria prepare themselves for cell division and multiplication.

Exponential or log phase: Exponential phase is a period of rapid cell division. Once the cell has accumulated all the necessary components for cell division, it starts diving exponentially. Microorganisms develop and divide at the fastest pace feasible given their genetic capacity, the type of medium, and the ambient conditions. The exponential or log phase of growth is characterized by predictable population doublings, where 1 cell becomes 2 cells, 4 cells become 8, and so on. During the exponential phase, their rate of growth is constant; that is, they complete the cell cycle and double in number at regular intervals. Situations that are optimum for the cells result in very rapid growth (and a steeper slope on the growth curve), whereas conditions that are less than ideal result in slower growth. Because cells at the exponential phase of growth are the healthiest and most uniform, they are used in the majority of experiments. This phase's length depends on bacterial circumstances.

Stationary phase: Throughout the exponential phase, organisms utilize the nutrients for their growth. Therefore, the nutrients in this closed culture system deplete at some point, waste products accumulate, and even space may run out, reducing cells. The growth curve flattens. In the stationary phase, the number of cells reproducing equals the number of cells dying. Some microorganisms produce secondary metabolites like antibiotics in this phase.

Death phase: The organisms cannot stay in the stationary phase infinitely. As conditions no longer encourage growth, the condition of the cells deteriorates, causing the growth curve to decrease.

9.9.1 Mathematics of Growth Kinetics:

When cultured in broth, microorganisms are typically incubated in a closed culture vessel with a single batch of media. Since no new medium is supplied during incubation, nutrient contents decrease, and waste concentrations rise over time. In batch culture, the population increase of microorganisms reproducing by binary fission can be represented as the logarithm of the number of viable cells vs. incubation time. Each microorganism divides at regular intervals during the exponential phase. As a result, the population doubles in size over a specific time period known as the generation (doubling) time (g). The term generation has the same significance for animals as it does for people. It is the time between an individual's birth and the time of producing offspring. Continually the population will double with the passing of each successive generation. A simple example will demonstrate this. Assume a culture tube contains one cell that divides every 20 minutes. Each new fission cycle or generation of bacteria doubles the population or multiplies it by a factor of 2. For example, after 20 minutes, the population of *Escherichia coli* will be 2 cells, 4 cells after 40 minutes, and so on. Because the population doubles every generation, population growth is always 2n, where n is the number of generations. The resulting population growth is exponential or logarithmic. This impact can continue at a steady rate so long as environmental conditions remain good.

Calculation of the growth rate constant

Let N_0 = the initial population number
N_t= the population at time t
n = the number of generations in time t
For populations reproducing by binary fission
$N_t = N_0 \times 2^n$
Solving for n, the number of generations, where all logarithms are to the base 10,
$\log N_t = \log N_0 + n \cdot \log 2$ and,
$n = \log N_t - \log N_0 / \log 2 = \log N_t - \log N_0 / 0.301$
The growth rate constant (k) is the number of generations per unit time (n/t). Thus
$k = n/t = \log N_t - \log N_0 / 0.301t$
Calculation of generation (doubling) time
If a population doubles, then
$N_t = 2N_0$
Substitute $2N_0$ into the growth rate constant equation and solve for
$k = \log(2N_0) - \log N_0 / 0.301g = \log 2 + \log N_0 - \log N_0 / 0.301g$
$k = 1/g$
The generation time is the reciprocal of the growth rate constant.
$g = 1/k$

Common problems with bacterial culture

Contamination - Contamination is the growth of unwanted microorganisms in a specific culture. A good aseptic technique can aid in the prevention of bacterial culture contamination.

Some species overgrow - When attempting to isolate a species from a mixed sample, these vigorous species may overgrow and obscure the presence of slower-growing target species. This can be mitigated by using selective media and optimal growth conditions for your target species (if known).

9.10 DIFFERENT FIELDS OF MICROBIOLOGY

Microbiology is a field of applied science that significantly impacts a wide variety of other fields, including genetics, biochemistry, food sciences, ecology, immunology, agriculture, medicine, and many more. Despite their relatively modest size, they provide the most valuable resource for biotechnology. Research on the genetics and molecular biology of microbes has been conducted using a wide variety of microbial taxa.

Pure Microbiology - Basic microbiology includes the study of the morphology, ecology, taxonomy, genetics, and physiology of particular microbe groups.

- Bacteriology - the study of bacteria
- Phycology, also known as the study of algae
- Mycology is the study of fungi, which includes yeasts and molds.
- Virology, sometimes known as the study of viruses.
- The study of protozoa is known as protozoology.

Medical Microbiology - The investigation of infectious bacteria, including their origins, life cycles, and physiologies. Medical microbiology is the study of genetics, pathogenicity, and control of microorganisms. Understanding how the immune system of vertebrates protects them against diseases and how they react to pathogens is an essential component of medical microbiology. This subfield is primarily concerned with investigating physical and cultural characteristics of resistance characteristics of bacteria, as well as methods for diagnosing, treating, and preventing infectious diseases.

Immunology - It is one of the fields that is expanding at the quickest rate and encompasses the causes of practical health issues and possible solutions. Immunology is the study of how a host responds to the presence of foreign material in their environment.

Environmental Microbiology - Environment microbiology is an integral part of microbiology that studies the role of bacteria in maintaining environmental quality. Since bacteria are present in every environment, including the air, water, soil, and food, they affect the degradation and decomposition of natural wastes (bioremediation) and the energy flow in ecosystems. The study also aids in comprehending the microorganisms of freshwater and saltwater. Recent research has demonstrated that some genetically engineered organisms can help in the cleanup of oil spills, hence enhancing the study of environmental microbiology.

Biotechnology - It is the most crucial branch that focuses on applying biological approaches for the benefit of humanity. It includes the utilization of microbes in manufacturing pharmaceuticals, fermentation of foods, and waste treatment. In addition, it covers the development of strategies for the production and manufacturing of particular molecules. It focuses on the nature of genetic information, the control, development, and function of a cell, and the way of producing new microbial cells using recombinant DNA technology, all of which apply to industrial microbiology.

9.11 APPLICATIONS OF MICROBIOLOGY

Microorganisms are recognized as vital research tools due to their contribution to comprehending the chemical and physical basis of life. As discussed earlier, they are the most prevalent group of living organisms in the biosphere and play an important role in our daily lives. Microbiology focuses mostly on analyzing the biochemical and genetic composition of living organisms. Since microbes are excellent models for studying cell functions and microbiology has applications in the health, agriculture, and cosmetics industries, microbiology is recognized as a crucial scientific field with vast promise. Let us examine the significance and uses of microbiology in greater depth.

Food Industry - In addition to edible fungi such as mushrooms, microorganisms such as yeasts, bacteria, cyanobacteria, and fungus are utilized as human food or animal feed. Microorganisms used in the production of cellulose or lignocellulose serve as food for humans as such or their products are utilized. Animal feed is also made from microbial products.

Energy derived from microorganisms - A variety of substrates have the potential to be converted into biogas by methanogenic bacteria, which can then be used as a source of energy. Microbes such as *Methanobacterium* and *Methanococcus* can use carbon dioxide (CO2) as an electron acceptor, which ultimately results in the production of methane. Ethanol can also be utilized in the creation of gasohol, which is created by combining 80% gasoline with 20% ethanol in the appropriate proportions. The oil content of microalgae presents an opportunity for use in the manufacturing of biodiesel. They also have a higher amount of lipids, which are the raw material used in the manufacturing of biodiesel, in their composition.

Microbes in Pharmaceutical Industry - Two major contributions of microbiology in the pharma industry are the discovery of antibiotics and the development of a vaccine. Recent advancements include phage therapy, which uses bacteriophages to treat bacterial infections. Steroids are another microorganism-produced pharmacological substance.

Production of Chemicals from Microorganisms - Using microorganisms results in the production of several different industrial chemical compounds. These compounds include lactic acid, ethanol, fructose, glycerol acetaldehyde, acetic acid, succinic acid, etc. The bacteria known as *acetic acid bacteria, lactic acid bacteria, propionic acid bacteria, butyric acid bacteria, E. coli,* and *Aerobacter aerogenes* are among the microorganisms that are responsible for the production of these compounds.

Microbiology's Importance in Biotechnology - Microorganisms are utilized in numerous biotechnology fields. In the fermentation sector, bacteria degrade complex organic materials to produce organic acids, fermented foods, ethanol, and vinegar. In molecular biology and recombinant DNA technology, microbes (e.g., viruses) serve as a source for molecular vectors such as plasmids, phagemids, and cosmids. Bioremediation eliminates organic chemicals and hydrocarbons from sewage water by decomposing these organic pollutants using microorganisms.

UNIT SUMMARY

This section gives an overview of the basic microbiological concepts and techniques. The initial part gives the idea about the concept of single-celled organisms. Further, the methods for the identification and classification of organisms are explained. Additionally, the significance of the microscope in the identification of these organisms has been discussed. Similarly, the ecological concept, media types, and uses, sterilization techniques have been emphasized. Further, the concept of bacterial cell culture has also been discussed.

EXERCISES

Multiple Choice Questions

1) A sub-type or a genetic variant of biological species is called
   A. Genus
   B. Strain
   C. Family
   D. Order

2) The time taken by an organism to adapt to new growth conditions is referred to as
   A. Log Phase
   B. Death Phase
   C. Lag Phase
   D. Stationary Phase
3) The temperature and pressure requirement for the autoclave is
   A. 10 lbs per sq. inch pressure, 121°C temperature
   B. 10 lbs per sq. inch pressure, 100°C temperature
   C. 15 lbs per sq. inch pressure, 121°C temperature
   D. 15 lbs per sq. inch pressure, 100°C temperature
4) A method used for surface sterilization of laminar airflow, biosafety cabinet is
   A. Infrared Radiation
   B. Ultraviolet Radiation
   C. Autoclaving
   D. Moist heat sterilization
5) Which physical form of media will you use to check the motility of a given organism?
   A. Solid media
   B. Semi-solid media
   C. Liquid media
   D. None of the above
6) Stuart medium is an example of
   A. Differential media
   B. Transport media
   C. Anaerobic media
   D. Selective media

Answers: 1) B; 2) A; 3) C; 4) B; 5) B; 6) B

Short Answer Type Questions

1) What is sterilization? Name the methods of sterilization.
2) Briefly explain the short streak plate method
3) Write a short note on the ecological aspects of the single-celled organism.

Long Answer Type Questions

1) Explain the pour plate technique and state its importance.
2) Briefly explain different phases of the bacterial growth curve
3) Briefly explain the different methods of sterilization.

4) Enlist different media used for microbial cultivation. Mention its differentiating factors along with examples.

NUMERICAL PROBLEMS

1) Bacterium divides every 35 minutes. If a culture containing 10^5 cells per mL is grown for 175 minutes, what will be the cell concentration per mL after 175 minutes?

Answer:

As we know, that bacterium divides after every 35 minutes through simple mitotic division; therefore number of divisions is 175 / 35 = 5

Since one bacterium on division produces two cells so cell concentration after 175 minutes will be

$=10^5 \times (2)^5$

$=32 \times 10^5$ cells/ml

PRACTICAL

1) Methods of preparation of glassware for sterilization.
2) Effect of UV light on the growth of microorganisms.
3) Preparation of growth media:
4) Liquid medium (Nutrient broth)
5) Solid medium (Nutrient agar)
6) Preparation of slants, butts, and plates
7) Inoculation of liquid and solid media
8) To plot a bacterial growth curve and determine
    a) Generation time and b) Specific growth rate of bacterial culture
9) Viable count: Spread plate and Pour plate

REFERENCES AND SUGGESTED READINGS

- Joanne, M. (2008). Prescott/Harley/Klein's Microbiology.
- Michael, J., Pelczar, J. R., Chan, E. C. S., & Noel, R. Krieg (2004). Microbiology.
- Salle, A. J. (1948). Fundamental principles of bacteriology. Fundamental Principles of Bacteriology., (Edn 3).
- Stanier, R. Y., Ingraham, J. L., Wheelis, M. L., & Painter, P. R. (1978). General microbiology 5 th edition.
- Kathleen, P. T. (2017). Foundations in Microbiology: basic principles. Mcgraw-Hill Us Higher Ed.

Dynamic QR Code for Further Reading

CO AND PO ATTAINMENT TABLE

Following is the blank CO and PO attainment table, one can use this for individual practice for CO-PO mapping. Course outcomes (COs) for this course can be mapped with the programme outcomes (POs) after the completion of the course and a correlation can be made for the attainment of POs to analyze the gap. After proper analysis of the gap in the attainment of POs necessary measures can be taken to overcome the gaps.

Course Outcomes	Expected Mapping with programme outcomes *(1: Weak Correlation; 2: Medium Correlation; 3: Strong Correlation)*												
	PO-1	PO-2	PO-3	PO-4	PO-5	PO-6	PO-6	PO-7	PO-8	PO-9	PO-10	PO-11	PO-12
CO-1													
CO-2													
CO-3													
CO-4													
CO-5													

INDEX

A

B

C

D

E

F

G

H

I

K

L

M

N

P

R

S

T

U

V